BOILER OPERATORS HANDBOOK

prepared by

National Industrial Fuel Efficiency Service Ltd.

Graham & Trotman

First published in 1959 as
the *New Stoker's Manual*
and in 1969 as *The Boiler Operators
Handbook*

This revised edition published in 1981 by
Graham and Trotman Limited
Sterling House
66 Wilton Road
London SW1V 1DE

Reprinted 1985

ISBN-13: 978-0-86010-251-9   e-ISBN-13: 978-94-010-9134-3
DOI: 10.1007/978-94-010-9134-3

Biddles Ltd, Guildford, Surrey

Typeset in Great Britain by
Input Typesetting Limited, London

CONTENTS

LIST OF FIGURES

FOREWORD

The popularity of the *Boiler Operators Handbook* has prompted the issue of a revised edition.

Other than a relatively small number of developments, essentially associated with solid fuel firing methods using the fluidised bed technique, no radical changes have occurred since the first edition of the Handbook was issued in 1969. In revising a work of this kind there is a great temptation to omit practices that are now less common in the UK. In view of the enormous pressure on Global energy resources, however, the chapters dealing in methods of hand-firing have been retained in the hope that they may be of value to those in the less developed nations where energy problems are infinitely greater than ours.

High combustion intensity boilers, commonly known as Package Boilers, of the Shell Construction design, have now much greater steam output than their predecessors and the need for high levels of maintenance and operating skills remain as essential as when this group of boilers first appeared on the market. Also the standard of water treatment required is probably higher than the Operator has been accustomed to.

The Health and Safety at Work Act re-emphasised the continued need for adherence to the principles that ensure a pressure vessel be maintained in a safe condition at all times. Accordingly the revised edition of the *Boiler Operators Handbook* has enlarged its sections on Safety and the Clean Air Act.

The *Boiler Operators Handbook*, like its predecessor the *New Stoker's Manual*, is intended to help the boiler operator carry out his important work with skill and efficiency. It is not a textbook, nor does it go into great detail. Within its compass however, it contains sufficient information to encourage the operator to study the subject more extensively and obtain a recognised qualification.

Any of the NIFES offices are prepared to discuss the availability of courses to enable the operator to further enhance his knowledge and skills.

Both the Imperial units and their SI equivalents have been included in this new issue.

Much inefficiency and many accidents are the result of ignorance and it is to be hoped that this new issue will help the operator and/or nominated attendant to understand the principles involved and enhance his interest in the plant under his care.

Chapter 1

FUELS IN COMMON USE

Fuel is any substance which can be burned to produce heat and all the common fuels are composed mainly of carbon and hydrogen. Wood, one of the earliest fuels, is still used today, mainly in the form of waste products from the timber trades. In this country, coal is the principal indigenous fuel source but large quantities of fuel oil are used for industrial, commercial, and domestic use. Nuclear energy is being used increasingly in the production of electricity and the discovery of reserves of natural gas has meant this fuel too is being used to fire boilers of all types.

The formation of coal, oil, and natural gas commenced millions of years ago when decayed vegetable matter was subjected to great pressure and heat. The types and properties of the fuels known today were, in fact, decided at that time by the variations which occurred in these processes.

COAL

The type of coal is decided by certain easily recognisable factors, the moisture, volatile matter, ash and fixed carbon contents, which are expressed as percentages of the whole and together constitute the proximate analysis of the coal. Volatile matter is driven off when the coal is heated and normally burns as a flame, when combined with oxygen.

Coals with a high percentage of volatiles tend to burn with smoke but anthracites and Welsh steam coals, with smaller amounts of volatiles and high percentages of fixed carbon, are unlikely to make smoke. The amount of volatile matter, therefore,

1

is one of the principal factors which determines the way that coal behaves when burned.

Typical proximate analyses for British fuels are:

Type of Fuel	Volatile matter %	Fixed Carbon %	Moisture %	Ash %
Peat	52	26	18	4
Bituminous Coal (1) [a]	34	51	6	9
Bituminous Coal (2) [b]	29	61	4	6
Anthracite	6	87	2	5

[a] Fairly typical, weakly caking coal.
[b] Typical coking coal.

Another important factor or property is the caking index of the coal, which indicates its tendency to cake on the grate and cause smoke when the mass is broken up. The use of a strongly-caking coal can prevent a boiler from steaming at its

Description of coal	Rank	Principal Source	Secondary Source
Anthracites	100	South Wales	Central Scotland
Dry Steam	201	South Wales	Kent
Coking steam	202–204	South Wales	Kent
Prime coking	301	South Wales, Durham	Kent
Very strongly caking	400	Durham, Yorkshire	Northumberland
Strongly caking	500	Yorkshire, Durham	Northumberland, Lancashire, Nottinghamshire North Derbyshire
Medium caking	600	Yorkshire, Nottinghamshire, Lancashire, North Derbyshire	Northumberland
Weakly caking	700	Yorkshire Nottinghamshire North Derbyshire Northumberland	Lancashire, North Staffordshire, Central Scotland
Very weakly caking	800	Nottinghamshire Yorkshire North Derbyshire Cannock Chase	Warwickshire Lancashire Lothians
Non-caking	900	Nottinghamshire North Derbyshire Leicestershire Fife	Warwickshire South Derbyshire

maximum designed load and this type of coal can also be difficult to burn on modern mechanical stokers.

Coals are generally classified by the National Coal Board ranking system, which takes account of the caking index, and those of medium and weakly caking properties, in ranks 600, 700 and 800 are amongst the best for steam raising.

CALORIFIC VALUE

The calorific or heating value of a fuel may be expressed in British Thermal Units per pound (Btu/lb), one Btu being the amount of heat required to raise the temperature of one pound of water by one degree Fahrenheit. The calorific value is only fully realised when one pound of fuel is completely burned. Typical values are 11 000 to 13 000 Btu/lb (25.586–32.099 MJ/kg) for bituminous coals and about 14 300 Btu/lb (33.26 MJ/kg) for anthracites.

SIZES

Coal is also classified by size. When mined, it is a mixture of sizes from large lumps down to dust and is graded by being passed over a screen, usually composed of a steel plate containing a number of holes of one size. For example, a screen with 3″ diameter holes lets through coal of less than 3″ in size and anything larger stays on it. Coal sizes are generally known by the following names:

Name of Coal	Passes through screen		Stays on screen size	
	in.	mm	in.	mm
Large Cobbles	6	150	3	75
Cobbles	4	100	2	50
Trebles	3	75	2	50
Doubles	2	50	1	25
Singles	1	25	½	13
Peas	½	13	¼	6
Grains	¼	6	⅛	3

Sized coal larger than singles is not often used in mechanically-fired boilers and many plants are fired by smalls, which consist of anything below a certain size down to dust; e.g. 1″ smalls pass through a screen with 1″ holes (25 mm).

Pulverised fuel (PF), solid fuel ground down to tiny particles, enables faster rates of burning to be obtained with solid fuel and is almost wholly confined to very large water-tube boilers.

It is of interest to compare PF with liquid fuel. Both solid and liquid fuels need to be 'conditioned' before being used in burner systems. Coal requires fine grinding, whilst the heavier grades of liquid fuels require their viscosity adjusting by preheating, to facilitate the production of a fine mist by the burner system.

Gaseous fuels are already in a conditioned state, that is, each particle is as small as it can be to facilitate its ready combination with oxygen in the air.

ASH

The ash content of the coal influences its calorific value and behaviour on a grate. A high ash content solid fuel reflects itself in a lower calorific value, whilst the lower the ash content, the higher the calorific value.

In the case of mechanical firing a compromise is usually arrived at, because the fuel must have sufficient ash to protect the grate and at the same time have a satisfactory calorific value.

An important characteristic of the ash from an operator's point of view is its ash fusion point or melting point and the resultant character of the fused product.

A high melting point ash does not result in heavy clinker unless the normal rate of burning of a coal is grossly exceeded.

A low melting point ash will certainly lead to operational problems and the extent will depend upon the type of clinker formed. A large, lumpy clinker formation can be broken with considerable effort, but a thin biscuit-like clinker which effectively covers the whole of the grate cannot readily be broken down and removed whilst the plant is in operation.

MOISTURE

The percentage of moisture in a fuel is reflected in the calorific value. A high moisture content in the fuel reduces its calorific value, and can also result in handling problems.

SULPHUR

The sulphur content of British Coals ranges from 0.7 to 2.8% and the problems associated with or as a result of its combustion to Sulphur Oxides are the same as experienced with Sulphur in Fuel Oils, i.e. mainly low temperature corrosion.

COKE

Coke is not often used in the Industrial and the larger Non-Industrial boiler plant these days, mainly because of the wide-spread use of Natural Gas and the closure of Gas Works.

Some coke from the high temperature carbonisation process at steelworks or from coke ovens supplying steelworks may still be used locally.

Recognition must be given to two important properties:

   (a) the abrasiveness of coke which can lead to severe main-tenance on mechanical handling systems; and
   (b) the reactivity of the fuel.

High temperature carbonisation processes produce a low reactivity coke reflecting itself in slow rate of change in its combustion when a peak demand has to be met.

The product of the low temperature carbonisation process used to produce household smokeless fuels has a high rate of reactivity and this coke is still used in Central Heating Plant where the rate of change in load is not rapid.

LIQUID FUELS

Liquid fuels are either of petroleum origin or are obtained from coal tar. Liquid fuels derived from the manufacture of coal gas are now no longer freely available since the closure of gas works throughout the country. Types of liquid fuels are classified by viscosity, or resistance to flow, generally measured as the time in seconds for a certain amount of the liquid at 100°F (37.8°C) to pass through a Redwood Viscometer.

The types of liquid fuels generally available together with some characteristics are shown in Table 1.

Flash point determination is a measure of the fire hazard when storing bulk supplies, e.g. Waste oils could have a low flash point due to the addition of paraffin and petrol.

GASEOUS FUELS

Gaseous fuels in general use are natural gas and liquid petroleum gases (LPG) and are specified by the Calorific Value expressed as Btu/cu ft/(Btu/ft^3) or as Megajoules per cubic metre (MJ/m^3) of the gas at normal temperature and pressure. Liquid Petroleum Gases (LPG) are stored in liquid state under pressure and must be reduced in pressure and vaporised before they can be burned with safety.

Town Gas, the product of high and low temperature carbonisation of coal, is now only used within the manufacturing area, e.g. steelworks, and any surplus is used in related industries associated with the plant.

Table 2 lists some properties of natural gas, LPG and town gas.

TABLE 1 TYPES AND CLASSIFICATIONS OF PETROLEUM FUEL OILS, WITH SOME CHARACTERISTICS

Grade of Liquid Fuel	Viscosity Seconds Redwood No.1 @ 100°F/37.8°C	Calorific Value Btu/lb. Gross	Calorific Value MJ/kg Gross	Specific Gravity	Heating Value Gall/Btu	Gallons Per Ton	Storage Temp. °C	Burner Temp. °C	Specific Gravity Correction per °F	Specific Gravity Correction per °C
Gas Oil	35	19 600	45.668	0.85	166 600	264	Atmospheric	Atmospheric	0.000 37	0.000 67
Medium Fuel Oil	200	18 900	44.037	0.93	175 800	240	8	43–60	0.000 36	0.000 65
Heavy Fuel Oil	950	18 750	43.683	0.95	178 100	236	21	77–94	0.000 35	0.000 63
Extra Heavy Fuel Oil	3500	18 300	42.639	0.98	173 300	229	40	110–121	0.000 35	0.000 63

Grade of Liquid Fuel	Air Required per lb Fuel lb	Air Required per kg. Fuel kg.	Waste Gas per lb. Fuel lb	Waste Gas per kg. Fuel kg.	Target O_2 in Waste Gases %	Target O_2 in Waste Gases %	% S	Flash Point Closed Min. °C	Mean Specific Heat between 0–100°C	Theoretical CO_2 in Waste Gases
Gas Oil	14.4	14.4	15.4	15.4	13	3.5	0.75	65.5	0.49	15.6
Medium Fuel Oil	14.1	14.1	14.9	14.9	13	3.6	3.2	65.5	0.45	15.7
Heavy Fuel Oil	14.1	14.1	14.9	14.9	13	3.8	3.5	65.5	0.45	15.9
Extra Heavy Fuel Oil	13.7	13.7	14.4	14.4	13	3.9	3.5	65.5	0.45	16.0

TABLE 2.   PROPERTIES OF TOWN GAS, NATURAL GAS AND LPG

Gas	Specific Gravity Compared with Air at stp	Flame Propagation Air/Gas ratio limits %	Calorific Value Btu/cu ft Gross	Calorific Value MJ/m³	Air for complete combustion Vol/Vol. Gas.[a]	Total Wet Products of combustion Vol/Vol. Gas.[a]	% theoretical CO_2 in Dry Waste Products	Target CO_2 in Dry Waste Products	Target O_2 in Dry Waste Products
Town Gas	0.45	5-30	375-550	13.98-20.45	4.5	5.0	13.0[b]	12.0	2.0
Natural Gas	0.593-0.602	5-15	1025-1049	38.16-39.05	9.5	10.5	11.7	10.0	3.2
Commercial Butane	1.90-2.10	1.8-9.0	3270	121.74	30.0	33.0	14.0	12.5	2.2
Commercial Propane	1.40-1.55	2.2-10.0	2500	93.07	24.0	25.8	13.7	12.2	2.3

[a] N.B. Vol/Vol. or m³/m³ or Cu. ft. /Cu. ft.
[b] Typical but varies with carbonisation process.

8

Chapter 2

COMBUSTION

Combustion is the name given to the chemical reaction between fuel and oxygen which results in the liberation of heat. The oxygen is usually derived from the air. A boiler operator should know enough about combustion to ensure that the full heating potential of the fuel is realised on his plant.

Combustion will not start in air or continue spontaneously below a certain temperature, known as the ignition temperature of the fuel in the air. Once the fuel is ignited, it must supply sufficient heat to raise the temperature and establish continuous combustion.

The air for combustion contains 21% by volume as oxygen and 79% as nitrogen. The nitrogen takes no part in the combustion process but passes unchanged through the furnace and so carries away heat.

Pure carbon can burn in air in two ways:

(i) If there is enough air, it burns to carbon dioxide (CO_2). The oxygen in the air is then replaced by a similar volume of CO_2. Thus, if the air is exactly right, the gaseous products of combustion contain 21% of CO_2 and 79% of nitrogen (see Fig. 1).

(ii) If there is too little air, some of the carbon burns to form carbon monoxide (CO). In this case, the full calorific value of the carbon is not released and combustion is said to be incomplete.

When hydrogen burns in air, the product of combustion is water. Initially, this takes the form of steam but, on cooling, condenses to water which occupies a negligible volume compared with the original hydrogen.

9

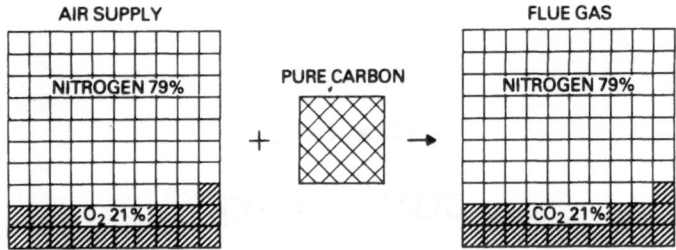

Fig. 1. Percentage of carbon dioxide in flue gas by volume

COMBUSTION IN PRACTICE

In fuels, carbon and hydrogen occur together and so burn together. When they burn together completely, the hydrogen combines with some of the 21% of oxygen in the air, to leave water and nitrogen, and the carbon combines with the remainder of the oxygen to form CO_2 and nitrogen.

If, therefore, exactly the right amount of air is provided to burn the fuel completely, the waste gases will contain less than 21% of CO_2. There are two possibilities: (a) the water in the combustion products will remain as steam or (b) it will condense. In both cases, the percentage of CO_2 in the products will be less than 21%, but will be lowest in (a) because of the volume occupied by steam.

When water is present as steam, analysis of the gases is on the wet basis. When the steam is condensed out of the gases, analysis is on the dry basis. As the waste gases are cooled and steam condensed in practically all types of analysis apparatus, the dry basis is the one generally used.

The first section of Fig. 2 shows what happens when a bituminous coal is burned with the exact (theoretical) amount of air to complete combustion and the steam is condensed out of the products of combustion.

There must be complete combustion in the boiler but practical difficulties make it impossible to supply the exact amount of air and excess air has to be introduced. Too little air results in the formation of CO, and probably of smoke, in the combustion products. Too much air is equally undesirable since it cools the

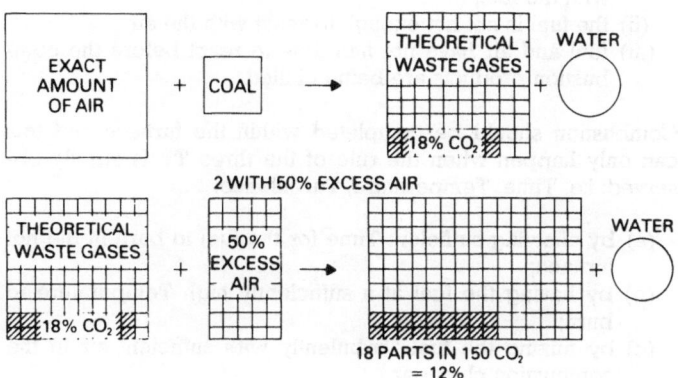

1 WITH THEORETICAL AMOUNT OF AIR

EXACT AMOUNT OF AIR + COAL → THEORETICAL WASTE GASES 18% CO_2 + WATER

2 WITH 50% EXCESS AIR

THEORETICAL WASTE GASES 18% CO_2 + 50% EXCESS AIR → 18 PARTS IN 150 CO_2 = 12% + WATER

Fig. 2. The combustion of coal

furnace and carries away useful heat. The second part of Fig. 2 shows the practical effect of burning bituminous coal with 50% excess air, with analysis of the waste gases on the dry basis.

The theoretical percentage of CO_2 in the waste gases, measured on the dry basis varies with different fuels but figures for the more common fuels, together with target figures for practical purposes, are given overleaf.

The figures are approximate but the operator should aim for the maximum figure obtainable with CO and with only a faint haze of smoke in the waste gases.

Fuel	Theoretical % CO_2 (Dry Basis)	Target CO_2 (Dry Basis)	Target O_2 (Dry Basis)
Bituminous coal	18.6	12.0	7.4
Dry steam coal	19.2	13.0	7.9
Coke	19.5	13.0	7.0
Anthracite	19.5	13.0	7.0
Fuel oil	16.0	13.0	3.9
Natural Gas	11.7	10.0	3.2

For one or more of the following reasons, it is possible in practice for combustion to be incomplete even though there is enough, or even too much, excess air:

11

(i) air passes through the furnace without mixing thoroughly with the fuel;

(ii) the fuel is not hot enough to react with the air;

(iii) fuel and air have not had time to react before the combustion products are being chilled.

Combustion should be completed within the furnace and this can only happen when the rule of the three T's is strictly observed: i.e. Time, Temperature, Turbulence.

(a) by allowing sufficient Time for the fuel to burn in the hot furnace;

(b) by having the fuel at a sufficiently high Temperature to burn;

(c) by mixing the fuel Turbulently with sufficient air in the combustion chamber.

In the absence of CO, the percentage of CO_2 in the waste gases indicates the amount of excess air used for combustion and it is essential for an instrument giving this information to be provided in the boiler house. If there is excess air in the waste gases, oxygen will also be present and an instrument which measures the oxygen percentage can be used instead of a CO_2 recorder. In this case, the aim must be to keep the oxygen percentage as low as possible while avoiding the presence of CO and smoke. The oxygen percentage gives a direct indication of the amount of excess air in the waste gases.

AIR SUPPLY IN PRACTICE

Since combustion is not instantaneous but takes place in stages, the fuel needs time to burn in the hot furnace with sufficient air to complete combustion. Air is, therefore, admitted in two ways: (i) as Primary Air entering the furnace with the fuel or, in the case of solid fuel burning on a grate, through the fuelbed; (ii) as Secondary Air admitted turbulently to complete combustion. Occasionally, tertiary air is also needed.

There are three general mechanisms by which combustion air can be supplied and sometimes a combination of the three is employed:

12

(i) **By Natural Draught** which is created when the hot gases pass up the chimney and cause suction in the furnace;

(ii) **By Induced Draught,** or fan-induced draught, caused by a fan located at the boiler outlet which sucks air through the system and augments the suction of the chimney;

(iii) **By Forced Draught,** or fan-forced draught, caused by using a fan located before the furnace to blow air through it.

Just sufficient air should be admitted to the furnace of a boiler to effect efficient combustion or useful heat will be carried away by the excess air and efficiency reduced. Air should, therefore, always be admitted under precise control and not allowed to leak into the boiler or flues through holes, cracks, or bare parts of the grate.

The amount of primary and secondary air is directly related to the fuel burning rate and the load on the boiler. Load variations call for alterations in the firing rate and air supply, therefore the primary and secondary air supplies have to be adjusted by altering the dampers, valves or louvres or, if a fan is used, possibly the speed of the fan. The relative proportions of primary and secondary air supply can thus be adjusted independently. Alteration to the setting of an individual damper will vary the draught, which is the difference in pressure between air or gases in the boiler or flue and the outside air. Draught is the motive force which makes the air or gases move and is usually indicated as negative or positive, on a draught gauge. The movement of air or gases is from the zone of higher pressure, towards the point of lower pressure, Draught is measured in inches wg or mbar.

Draught gauges are usually connected to points just after the forced draught fan damper, in the furnace, in boiler flues before the induced fan or chimney damper if there is no induced draught fan, and at the base of the chimney. The draught reading should always be used as a guide to the effect of opening or closing draught controls or dampers. The correct draught is that which keeps steam pressure steady whilst maintaining the highest CO_2 and the lowest percentage of oxygen in the waste gases and preventing the presence of smoke and carbon monoxide. Some furnaces operate under pressure and others under suction but the flow of gases may be similar in each because it is the pressure difference between two points in a system and not the absolute pressure that determines the flow of the gases.

13

COMBUSTION OF COAL ON A GRATE

Figure 3 illustrates the principle of coal burning on a static grate. There are very few hand-fired plants in use today, what there are being mainly confined to small central heating units but the principles shown hereafter remain the same for mechanically operated grates. Mechanical grates have usually been developed from well-tried manual firing techniques, and an appreciation of this is essential in order to fully understand mechanical firing.

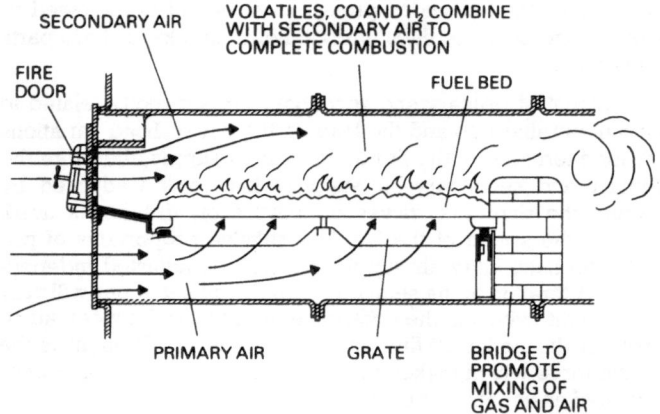

Fig. 3. Coal burning on a grate

Primary air is supplied beneath the grate and as it passes through the coked fuelbed, combines with the carbon in the coal and is converted into CO_2 and CO. These gases, together with the unchanged nitrogen in the air, rise from the fuelbed and mix with the volatile vapours. At this point, secondary air must be admitted turbulently while the gases are hot enough to ensure complete combustion of the volatiles and CO in the furnace. Secondary air is admitted through an adjustable grid in the furnace door or through forced draught ports which can be adjusted by a damper in the air supply duct leading from the forced draught fan.

Fresh air is raised to ignition temperature by the coal burning on the grate. The rate of combustion is varied by altering the

14

primary air supply and making a proportional adjustment to the secondary air in accordance with the draught gauge reading.

BURNING OIL, GAS AND PULVERISED FUEL

These fuels are not burned on a grate but by spraying or blowing into the furnace. The volume must, therefore, be adequate for the rate of firing and of the correct shape. Oil and pulverised fuel burn as fine particles projected into the furnace with the combustion air; proper mixing of air with the combustibles is much easier with these fuels than when burning solid fuel on a grate. Considerably less excess air is needed, usually less than half of that required for burning coal.

The rule of Time, Temperature, Turbulence has already been discussed and fuel which is only partially burned must neither be chilled by impinging on cool surfaces nor leave the furnace zone before combustion is complete otherwise smoke and un-burnt gases, e.g. carbon monoxide (CO), will result. Care must be taken to see that the burner is properly aligned in relation to the furnace tube or combustion chamber.

Pulverised fuel, in the form of fine particles dispersed in primary air, is supplied through a burner, secondary air being provided under pressure around the burner in a way which ensures turbulent mixing.

An oil burner atomises the oil into a fine spray. Primary air enters the furnace with the atomised fuel and secondary air is provided around the burner. Alternatively, all the combustion air can be supplied turbulently around the burner.

With Natural Gas burners normally all the air is supplied around the flame; alternatively primary air may be mixed with the gas (pre-mix burners) and secondary air is supplied around the burner.

In some of the light oil burners, the oil is vaporised by heat before burning and combustion is similar to the burning of a gas.

15

FLAMES

Flames are formed when fine particles of solid fuel, droplets of liquid fuel, gas, or the volatiles from coal, are mixed with air and ignited. The flames will, however, only persist and propagate through the mixture between certain upper and lower limits of the fuel/air ratios. These are known as the upper and lower limits of flammability; they vary widely from fuel to fuel and with certain conditions, such as temperature and pressure.

If such a mixture between the limits is contained in a long tube the flame will travel along it at a definite speed. This speed varies from fuel to fuel, with the air/fuel ratio and physical conditions such as pressure, temperature, the diameter of the tube and the direction, upwards or downwards, of flame travel.

Flame speeds are important in burners particularly where part of the combustion air, is pre-mixed with the fuel, and passes into the furnace via a pipe. If the flame speed is greater than the speed of the mixture in the supply pipe, the flame can burn back or flash back down the pipe, as sometimes happens on a domestic gas cooker. If, however, the flame speed is much less than that of the mixture issuing from the burner, the flame will detach from the burner nozzle and lift off. These effects, commonly called backfiring or light-back and lift-off or blow-off, are the reasons why burners for town gas cannot be used for natural gas.

Air/fuel mixtures can accumulate in a furnace, flue, boiler house, or other restricted space, and cause an explosion if ignited. Usually, the more restricted the space the greater the effect of the explosion. If pulverised fuel, oil or gas burners lose ignition, extreme care must be taken to ensure that the whole system is thoroughly purged of combustible mixtures before an attempt is made to relight the burner. In some automatic systems, flame failure devices are linked with controlled purging.

FLAME TYPE

Flames may be luminous or non-luminous. Oil and coal produce luminous flames, whilst Natural Gas burns with a non-luminous flame. Luminous flames have greater radiating powers when compared with non-luminous flames, e.g. on conversion from oil to Natural Gas less heat is transmitted in the furnace tube resulting in a higher flue gas temperature in the region of the tube plate. Overheating and cracking may be the consequence.

Chapter 3

WATER AND STEAM

STEAM RAISING

At atmospheric pressure, water boils at 212°F (100°C). The boiling point rises when the pressure of water vapour over the surface is raised above atmospheric pressure. When the pressure is reduced, the boiling point is lowered. For example, when water is heated inside the closed space of a boiler and the pressure is raised to 100 lb/in² (6.9 bar), it does not boil until the temperature reaches 337.9°F (169.9°C), nearly 126°F (69.9°C) above boiling point at atmospheric pressure.

If heat is added to something we expect it to become hotter! i.e. it makes sense! that is why it is called Sensible Heat. **Sensible heat** can be added to or removed from any substance without changing its physical state. The amount of sensible heat increases as the temperature rises.

When heat is added and the pressure is kept constant, a liquid will boil and some of it will change into vapour; this change in physical state takes place without alteration in temperature. The vapour produced from boiling water is steam and the amount of heat needed to change water into steam at the same temperature is known as the **latent heat** of evaporation. Latent heat is not constant and is reduced as the pressure of steam above the water increases. The latent heat of steam at atmospheric pressure is 970.6 Btu/lb (2257.5 kJ/kg). At 100 lb/in² (6.9 bar) it is 881.6 Btu/lb (2050.5 kJ/kg).

One pound of water at 32°F (0°C), can produce one pound of steam if it is raised to boiling point 212°F (100°C) by the addition of sensible heat and evaporated by the addition of latent heat. The sum of these two additions of heat is known as the **total**

18

heat, or enthalpy, of steam. While the value of the latent heat becomes less as the pressure rises, the value of the sensible heat increases, so there is a steady increase in total heat up to pressures of around 400 lb/in² (27.6 bar), after which the total heat begins to fall off. When steam is formed and all the water has evaporated, the addition of heat will raise the steam temperature. This further increase in heat is known as the superheat of steam. It is emphasised that superheaters must be fitted externally to the boiler or water container in order to raise the steam temperature at any given steam pressure.

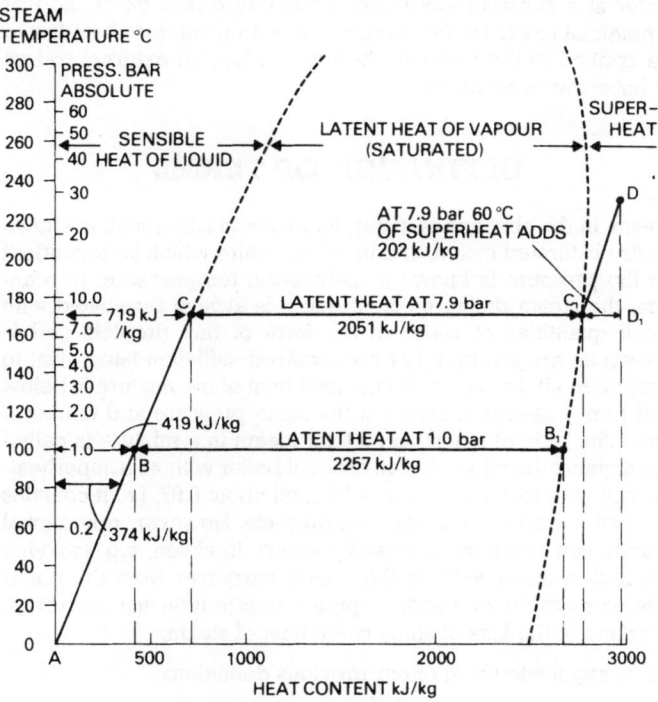

Fig. 4. Properties of steam

The facts given above are shown in Fig. 4. The horizontal straight lines are shown as additions of latent heat at atmospheric pressure and 100 lb/in², respectively, whilst the sloping line on the left of the diagram shows the rise in water temperature with

19

pressure and the addition of sensible heat. The sloping line C_1D, shows the added superheat temperature and C_1D_1 the gain in heat due to superheating by 60° at 7.9 bar.

If hot water is available to the boiler at 180°F, (82.2°C) from condensate returns, only 161 Btu/lb (374 kJ/kg) would be needed to raise the temperature to its boiling point at 100 lb/in², (6.9 bar.g) whereas if no such supply were available 309 Btu/lb (719kJ/kg) would be required, assuming that the initial water temperature is at freezing point; which in practice would be most undesirable. However, there are many boilers using feed water at a considerably lower temperature than 82.2°C so that a potential exists for fuel saving, since so much less heat has to be applied to the water in the boiler when an external source of hot water is available.

DEFINITION OF TERMS

Steam, in the absence of water, and without superheat, is known as **dry** saturated steam, and its temperature which is dependent on the pressure is known as **saturation temperature**. In practice, the steam delivered by a boiler is always associated with small quantities of water in the form of tiny droplets and is known as wet steam. It has not received sufficient latent heat to evaporate all the water and the total heat of the mixture is below that of dry saturated steam at the same pressure and temperature. The ratio of dry steam to wet steam in a mixture is called the **dryness fraction**. A conventional boiler without a superheater may give steam of dryness fraction about 0.97, i.e. it contains 3.0% of unevaporated water as droplets. However, operational factors can result in increased wetness in steam, e.g. carrying too high a water level in the boiler, carryover from the boiler due to **priming** or because pipe condensation has occurred, resulting in the loss of some latent heat of steam.

To re-capitulate some of our previous definitions:

Sensible Heat	Produces a rise or fall in temperature
Latent Heat	Produces a change of state, i.e. liquid to gas, ice to water, water to vapour or steam in the case of a boiler, without change in temperature.
Total Heat	Sensible Heat + Latent Heat.

A further property of steam has to be mentioned, namely its density and volume. The volume/lb of steam decreases with increase in pressure, and hence its density increases, a very important consideration when designing steam distribution systems. Steam at low pressure requires larger steam mains than does the same quantity at higher steam pressures.

The properties of steam are contained in Steam Tables generally under the following headings.

COLUMNS								
ABS.Pressure and/or Gauge pressure	Temp. (T)	Absolute Temp.	Sensible heat	Latent heat	Total heat	Volume Water	Volume Vapour	Density

WATER

Drinking and Industrial water is impure in the sense that it contains a variety of salts or chemicals and dissolved gases, such as oxygen and carbon dioxide, in solution. The composition of water will depend upon its source and geographical location. For example around the London Area the water contains a great deal of Calcium and Magnesium Bicarbonate in solution and the water when boiled leaves a heavy deposit of Calcium and Magnesium Carbonate behind. Further it does not readily produce a lather when soap is added to the water, and it is described as Hard. This is because the water is obtained largely from underground strata which is high in Calcium and Magnesium salts.

Mountain waters on the other hand being surface drainage waters are virtually rain water and as such are very soft and very little deposition occurs even after repeated boiling and the water lathers readily with only small soap additions. The kind of Hardness that can be removed by boiling is known as Temporary Hardness.

Water also contains other salts which cannot be removed by boiling but in saturated solutions are deposited on boiler surface and by their crystalline nature form very hard scale. These hardness forming salts can only be removed by chemical means and they are given the general designation Permanent Hardness. The combination of Temporary and Permanent Hardness together make up the Total Hardness of a water.

21

Water also contains salts which do not result in Hardness but together they all result in an overall category best described as Total Dissolved Solids. However, both within the grouping of salts described as Permanent Hardness and of salts which do not result in hardness there are some which can corrode steel surfaces. Therefore it can be seen that a boiler plate covered with water can be subjected to scaling and corrosion.

The aim of water treatment is to render both the scale and corrosion bearing salts to a relatively harmless state.

Even with good quality feed water, some impurities are left behind by heating and constant evaporation into steam and the boiler water becomes progressively contaminated by a concentration of dissolved salts, solid particles and sludge which can lead to priming. Continuous and intermittent blowdown valves for this purpose are provided on the boiler. The amount of blowdown is determined by chemical analysis or the specific gravity of a water sample drawn from the boiler. This specific gravity is directly related to the concentration of salts in the water and is expressed as Dissolved solids/parts per million or ppm.

PRIMING

This is caused by a vigorous surging of the water and such a condition can be caused by chemical contamination of the water or by mechanical and operational factors.

CHEMICAL CAUSES

(a) Organic matter in the boiler water
(b) Suspended solids in the boiler water
(c) Presence of oil in boiler water
(d) Soaplike substances in boiler water
(e) High alkalinity

MECHANICAL CAUSES

(a) Operating at a steam pressure below the design pressure, which results in an increased steam velocity from the water surface at any one output.

(b) Too high a water level, which will encourage priming especially when the load changes from low to high.

(c) Incorrect firing rate relative to the load requirement.

(d) Operation of on-off feed pump in relation to rapidly fluctuating load.

(e) Defects in internal design of boiler shell.

BUMPING

Rapid changes in water level can be observed in the gauge glass and are very similar to that arising from foaming.

Bumping may occur at very steady load conditions and is caused by a lag in the formation of steam bubbles. Superheating of the water at various point sources can occur accompanied by violent bumping.

FOAMING

Because of the various chemicals and solids in the water, and their effect on surface tension, a layer of froth or stable foam can build up on the surface of the water. The foam thickness is variable but can fill the steam space. There need not necessarily be any movement in the gauge glass when foaming occurs, however, foaming and priming can occur together. Perhaps foaming is best illustrated by reference to boiling peas in a lidded saucepan, a foam results, the lid lifts and we have boiling over—in more ways than one—if the male member has been told to keep a sharp eye on the pot and is listening to the radio instead.

This little analogy also serves as a sharp reminder that constant vigilance is required when boiling or looking after a boiler.

CARRY-OVER

This is the result of water being carried into the distribution system. This water carries soluble salts which can deposit in engine cylinders, turbine blades, valve bodies, traps etc. and result in serious damage to the plant.

Carry-over can be deleterious in many industries and, for example, can affect dyeing operations since the salts carried over will very often be of alkaline type.

The first essential in preventing these occurrences is to apply correct boiler and feedwater treatment and to follow the advice of the specialist or Company Chemist carefully.

Operational factors must also be considered:

1. Open stop valve and process valves gradually.
2. Adhere to the lowest possible water level without endangering safety.
3. If a Company's own water sources are used it must be adequately filtered to reduce any suspended matter to a low level.
4. See that no areas of boiler scale are left on the surface of the steel plate, and examine the raw water for traces of oil.
5. Anti-priming baffles should be correctly fitted and maintained thereafter.
6. The continued use of overloaded boilers should not be encouraged.
7. See that heat exchanger systems do not leak any 'foreign' material into the boiler feedwater system.
8. See the gauge glasses are properly aligned.

EFFECTS OF SCALE

Total hardness in the water, if not counteracted by correct treatment procedures, will result in scale formation. A layer of scale on the metal surfaces of the boiler acts as an insulator and will reduce the rate at which heat is transferred from the hot zones to the water. The conductivity of a calcium sulphate scale is about one hundredth that of steel! A layer of scale as thick as the wall of a boiler tube will result in a resistance to heat flow considerably higher than the bare tube. In practice, taking into account the film resistance on the water and gas-side, the effect of such a scale would be to lower heat transfer by some 20%.

Two results will follow:

1. To maintain boiler output at the acceptable level more fuel has to be used.
2. Since the heat transfer is retarded the metal becomes increasingly hotter to a point at which it becomes deformed and even ruptures with disastrous consequences.

CORROSION

Natural water in contact with the atmosphere dissolves air, of which the most active constituents are Oxygen and Carbon Dioxide. Their effect is seen daily in the active rusting of ferrous metals, tools which are not regularly oiled, steel windows which are not painted.

The effect of oxygen in a boiler is electro-chemical and differences in local concentration can produce electrical potential differences.

Carbon dioxide in water is a weak acid and soft water could well be slightly acid and therefore conducive to corroding ferrous metals. Carbon dioxide is also present in water as bicarbonate and carbonate and the former in particular will decompose and liberate carbon dioxide into the steam mains and subsequently condensate lines which could become corroded due to the acidity imparted to the water.

The removal of dissolved oxygen is an effective way of preventing corrosion and that can be achieved in either one or both of two ways, namely by mechanical de-aeration or by chemical means.

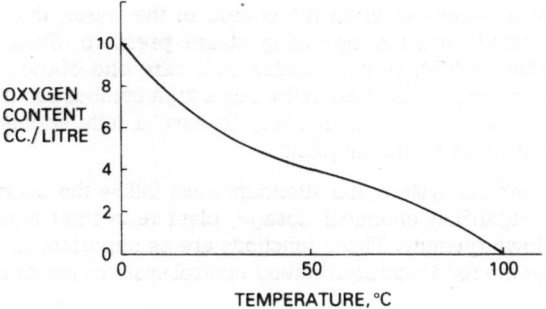

The former ejects the oxygen by heating the water to a point where it is liberated from the water before entering the boiler. The reader must have observed when water is heated that minute bubbles are progressively formed before the larger steam bubbles are produced by actual boiling.

The above diagram shows how the dissolved oxygen content of water is reduced as the water temperature is increased.

Chemical de-oxygenation, is achieved by the continuous addition of a chemical, such as sodium sulphite which combines with oxygen to form sodium sulphate.

Since raising the temperature of water reduces its oxygen content, the higher the feedwater temperature the better, regardless of whether or not a de-aerator is used. Less chemical addition will be required to fix the oxygen thus there could be a dramatic fall in the dissolved solids concentration. A lower concentration of dissolved solids means less blowdown, and less fuel wastage.

Hot condensate return has a two-fold effect since it directly saves fuel and contains less oxygen than a feedwater containing a high proportion of make-up water.

The boiler attendant must be on the alert to see that oxygen cannot enter the feedwater distribution system. There is little point in adding chemical to fix oxygen if an inspection cover is left open thus exposing the water to air.

WATER TREATMENT SYSTEMS

It is not within the scope of this Handbook to describe in detail the various systems of water treatment available. Their complexity will depend upon the source of the water, the type of boiler used, and the operating steam pressure. Since boiler attendants will likely have under their care one of the modern 'Package' boilers, best described as a high combustion intensity boiler they are more than likely to have a fully sophisticated external water treatment plant.

Whatever the system, the attendant must follow the instructions given regarding chemical dosage, plant re-cycling times, and blowdown quantity. These functions are as important as checking that the feed pumps and feed control systems are operating.

BLOWDOWN

The principal method of maintaining the total dissolved solids at a figure recommended by the boiler manufacturer is to resort to blowdown where some of the concentrated water and sludge is removed from the boiler by means of a valve. The extent of the removal is determined beforehand by analysis or by specific gravity measurement.

Manual intermittent, continuous blowdown or automatic blow-down systems can be used and where instructions are issued by the boiler manufacturer these should be studied carefully since blowdown can sometimes seriously interfere with water circulation over the heating surfaces and cause damage. A routine blowdown procedure should be established and maintained.

Where there is continuous blowdown, a small amount of water constantly flows away and it is not necessary to raise the water level before blowing down. The blowdown rate may be insufficient to clear the solid deposits at the bottom of the boiler and it is usual practice to supplement continuous blowdown with intermittent hand blowdown to a set amount.

The quantity of blowdown, once set, should not be altered, unless other instructions are given following a water analysis. The flow from continuous blowdown is often passed through a small heat exchanger for preheating the feed.

Automatic blowdown can also be applied and in that case, blowdown is effected by means of an automatic valve which opens for a predetermined time interval based upon the water analysis. Flash steam from the blowdown is quite often used for pre-heating the feedwater.

Boiler attendants should become conversant with the test required to determine the amount of blowdown, under the guidance of the visiting feedwater treatment specialists.

The simplest method is to use a salinometer to determine the specific gravity of the boiler water.

The apparatus is similar to that used in testing the specific gravity of acid in a car battery or of the specific gravity of home made wines and beers. The test is made at near atmospheric temperature and the reading then corrected to 15.5°C. The reading is approximately related to the total dissolved solids (TDS) in the water and reference to tables enables the amount of solids to be read off. The portable conductivity meter is also a quick simple way of measuring TDS.

The approximate amounts of dissolved solids for various types of boiler are generally as follows, but follow boiler manufacturers instructions.

Type of Boiler	TDS ppm
Lancashire Cornish	10000
Economic: two pass	4500
three pass	3500
Package	2000
Vertical	4500
Water tube	1000 at high pressure 5000 at low pressure
Locomotive	3500

In general, the higher the steam pressure, the lower the acceptable solids figure and this is also true of the 'Packaged' boiler or boilers which have a high rate of heat transfer.

Superheat temperatures will fall in those boilers fitted with superheaters if priming occurs due to a high concentration of dissolved solids in the boiler water.

Chapter 4

BOILER EFFICIENCY AND HEAT TRANSFER

BOILER THERMAL EFFICIENCY

It may be that some boiler attendants, especially those whose main field of work is not necessarily the care of a boiler plant, are not aware of the meaning of the term thermal efficiency. As applied to boilers it is simply the ratio of the Useful heat output in steam to the heat input in the fuel

Thermal Efficiency =

Heat in Steam above
    Feed Temperature (kJ/kg) × wt of steam/kg fuel

Calorific value of fuel kJ/kg

The heat in the steam can be obtained from steam tables, whilst the heat in the fuel must be determined in the laboratory. In the case of petroleum fuel oils and natural gas, the published data for their calorific or heat value is sufficiently constant and accurate for most practical purposes.

In order to arrive at the ratio, the weight of fuel used and the total steam output from the boiler must be known.

The thermal efficiency may be expressed on the gross or the net calorific value of the fuel.

29

GROSS OR HIGHER CALORIFIC VALUE

All fuels contain some hydrogen. When hydrogen is burnt it combines with oxygen to form water vapour, which normally leaves the boiler as steam in the flue gases at exit gas temperature. However, when determining calorific value in the laboratory, the products of combustion are allowed to cool to the original temperature of the fuel before combustion occurred. That is, the steam gives up its heat and condenses to water, which itself is allowed to cool. All the potential heat in the fuel is therefore accounted for and is known as the Gross Calorific Value.

In practice, the flue gases are rarely cooled to a point where the latent heat in the steam formed by the combustion hydrogen is recovered by the formation of water, that is, not all the heat in the fuel is recoverable and this results in what is termed the Lower Calorific Value or Net Calorific Value.

A parcel has sometimes marked upon it the Gross Weight, which is the weight of the Packaging Materials and the Content. The Net weight refers to the Contents only and since one is only interested in the contents, the packaging material is discarded. For any given boiler, the attendant may only be interested in the heat which that plant can recover and for this reason, the Net Calorific Value may be more meaningful to him, although many boilermakers quote efficiency on Gross Basis.

The thermal efficiency of a boiler depends on many factors, including the type of boiler, the quality of combustion, the load, the pressure and the general condition of the boiler particularly with regard to cleanliness.

Having defined thermal efficiency, it must be stated that boilers do not operate at 100% thermal efficiency and that there are heat losses associated with the operation of boilers. The source of these heat losses is the same, regardless of boiler design but their basic magnitude is different dependent upon boiler design. The magnitude of the losses can vary for any given type of boiler dependent upon the skill of the attendant, the excellence or otherwise of the burner, or mechanical stoker and the controls and the state of the maintenance of the boiler. The load condition and quality of the fuel, solid fuel especially, exerts an influence on the degree of heat loss likely to be present.

It is more important for an attendant to know the source and magnitude of the heat losses rather than the absolute thermal efficiency, since the extent of these heat losses are under his direct control and if these heat losses are kept to a minimal value then it follows that the thermal efficiency of the plant in his care will be at the highest possible level.

HEAT LOSSES

%
___

Heat loss in flue gases

Heat loss due to unburned fuel in
all residue (solid fuel firing)

Heat loss due to radiation and
other, unmeasured, losses.

___

Total loss

___

Thermal Efficiency = 100 −heat losses

HEAT LOST IN FLUE GASES

This should be regarded as the greatest single heat thief in the operation of boiler plant.

To some extent this has already been discussed in Chapter 2 where the influence of excess air and its relativity to CO_2% or free oxygen % in the flue gases was defined.

One important factor linked with excess air was not discussed at that stage, namely flue gas temperature. Before discussing the effect of varying temperature on flue gas loss it is desirable to see the effect of varying CO_2 at a fixed temperature on such a loss.

Whilst it is not the intention in a handbook of this character to become involved in mathematical expressions, there is a very simple formula based on the net calorific value of the fuel, which relates stack losses to variations in CO_2 and temperature, namely:

Sensible heat in flue gas $= \dfrac{K(T_1 - T_u)}{CO_2}$

where CO_2 is percentage CO_2 at boiler (econ) outlet;
T_1 is the temperature of the flue gases leaving the boiler (or economiser) °C;
T_U is the temperature of the air entering the boiler furnace °C; and
K is a factor dependent upon the fuel:

Fuel Oil	0.56	Natural Gas	0.38
Anthracite	0.68	Coke	0.7
Bituminous coal	0.63		

For further details see BS845:1972.

Example:

$$T_1 = 220°C$$
$$T_u = 20°C$$
$$CO_2 = 7.0\%$$
Fuel oil $= K = 0.56$

Stack Loss $\quad \dfrac{0.56\,(220{-}20)}{7.0} = 16\%$

Compare the same plant operating at 13.0%

Stack loss $\quad \dfrac{0.56\,(220{-}20)}{13.0} = 8.6\%$

It can be seen that reducing the excess air, as shown by operating the plant at a higher CO_2 halves the flue gas loss in this example, and the thermal efficiency of the plant has been increased by 7.4%

EFFECT OF RAISING FLUE GAS TEMPERATURE

Assume that T_1 is now 300°C and CO_2 13.0% as before.

Stack Loss $\quad \dfrac{0.56(300-20)}{13.0} = 12.1\%$

When compared with the example above, the heat lost in the flue gases has increased from 8.6 to 12.1% or 3.5 units, i.e. a loss in efficiency of 1% for every 22.8°C rise in flue temperature.

At boiler efficiencies around 75–80% the increase in fuel consumption is approximately 1% for every 18°C rise in flue gas temperature.

Clearly some degree of instrumentation is desirable to enable an attendant to operate his boiler plant efficiently and this will be discussed in a later chapter. If the Management have not had a quick look at this chapter before handing the book on to the Boiler Attendant, perhaps the attendant could tactfully show the significance of the above examples to the Management and perhaps they in turn will then provide the basic instrument to monitor combustion or have simple tests carried out by specialist firms from time to time.

The Lancashire boiler setting and many of the older water tube boiler walls are constructed of brick and the boiler operator must ensure that the brickwork is free from holes, cracks and other sources of air leaks, which reduce the draught and allow cold air to enter the plant. A regular check using a flame to detect air leaks should be made and the leaks sealed without delay.

An over capacity burner on an oil or gas fired boiler can, in some circumstances cause periodic cut-outs of firing. During these shutdown periods, cold air can enter through the furnace and cause loss of heat and efficiency. Where solid fuel is fired, underfeed and gravity feed stokers particularly are subject to the same losses as oil and gas burners if the thermostat or pressurestat causes the heat input to be cut off. It is important, therefore, to adjust the fuel input rate so as to keep the burner or stoker operation as continuous as possible.

Nevertheless, without basic instrumentation the operator must be aware of the areas in which changes in a preset or design condition can occur.

AIR SUPPLY

This has already been discussed in Chapter 2 but the attendant must be alert to visual change regardless of the type of fuel used. For example, if smoke can be seen, is the fuel bed too thick (when solid fuel is used) or have the primary and secondary air dampers been correctly adjusted? This latter condition can be common to all fuels. If butterfly dampers are used, do they move freely, have they become jammed in the partially closed mode, or even in the wide open mode? Are the fans rotating in the right direction?

If liquid fuel or natural gas is the fuel used, has the flame length altered, has the noise level of the burner changed? Inspect linkages between fuel and air supply — are they moving correctly?

Sufficient has been said to indicate that constant vigilance is necessary and that a casual glance at the pressure gauge by a man designated to visit the boiler house is not enough. If a casual attendant is employed, let not his approach be casual.

FLUE GAS TEMPERATURE

Its significance on the magnitude of stack losses has already been demonstrated, but how do changes occur and has the boiler attendant any degree of control over these changes?

DIRTY HEAT TRANSFER SURFACES

Anything that can impede the transfer of heat from the fuel bed, the flame, hot flue gases through the metal plate or metal tube into the water increases the final flue gas temperature. That is, soot and fly ash on the flue gas side of the metal and scale on the water side of the metal act as effective barriers to heat flow.

The most effective guide to the rate of build-up is undoubtedly the measured increase in the final flue gas temperature.

In the absence of instrumentation, examination of the heat transfer surfaces on the flue gas side should be carried out regularly and a schedule of cleaning arranged. On the water side, having instituted treatment systems, regular testing of both raw and boiler water should ensure a high degree of cleanliness.

In those boilers which have brickwork baffles to separate various gas passes, such baffles should be examined to see that there is no short-circuiting of the flue gases. The hot gases must 'scrub' every part of the metal at each stage of its travel, otherwise heat transfer will be impaired and the flue gas temperature will be higher than is necessary.

Again, inspection is recommended in the absence of instrumentation.

In the case of liquid fuel and natural gas burners, delayed mixing of air and fuel, resulting in long flames can increase flue gas temperature. Inspect fuel/air ratio system linkages, burner nozzle diameters, avoid unnecessary tampering with fuel/air ratio, do not enlarge oil burner jets, nor increase fuel pressures above the design requirement, since all or one of these can affect the final flue gas temperature. The foregoing are just a few examples relative to the operating factors which can influence flue gas temperature.

BOILER LOAD

It is easy to observe load changes where a steam meter is installed, but the fall and rise of the steam pressure gauge, together with the rate of firing also gives an indication of load. If the flue gas temperature is measured the attendant will soon be able to relate his measured temperature with that of load and recognise whether an increase can be attributed only to load changes and not to build up of deposits. The latter will be seen as a sustained and progressive increase with time, whilst a sustained temperature at different load levels will indicate the effect of load changes. In some instances, where a boiler is overloaded, deliberate overfiring is practised by either increasing jet size and/or increasing fuel oil pressure, or by increasing the rate of flow of fuel gas. Such a practice results in increased flue gas temperature but can manifest itself fairly rapidly in increased maintenance costs, e.g. rapid brick quarl failure, 'throwback' of heat on to the burner and burner front, cracking of fire tube ends in the Economic type boiler, damage of tube plate etc. With coal fired systems it can show up as excessive maintenance of grate bars or links, excessive amount of un-burned fuel in the ash with an additional increase in heat loss.

If a second boiler is available, the attendant should be able to recognise the need, and avoid sustained overload conditions.

CARBON IN ASH

Fuel beds should be disturbed as little as possible compatible with the breaking up of clinker and masses of coke, together with the removal of ash. Some carbon in ash is inevitable but time must be given for the fuel to be burned before ash removal is effected. Similarly, a moving grate running too fast or with too thick a fuel bed will result in excessive carryover of unburned fuel. Fuel which is continuing to burn in areas which have not been designed to bear excessive heat can result in a short life span for that particular section of the handling plant. It is with a solid fuel firing system that the skill needed by the attendant is greatest.

HEAT LOST IN DUST AND FLY ASH
(Grit loss)

In order to keep the carbon in dust at a relatively low level, care must be taken in adjusting draught levels. Excessive forced draught will lift fine particles from the fuel bed where they are entrained by the flue gases and carried forward into the various passes of the boiler and some then rejected via the chimney.

The physical size of the fuel influences the degree of carryover, but the operator must ensure that his primary air pressure is relative to the load required and that his secondary air pressure is relative to the degree of turbulence required to complete combustion. The draught must be only sufficient to overcome the resistances of the various flue gas passes and must not be so high as to accentuate the forward movement of the particles, thereby preventing them from falling back into the combustion zones.

RADIATION AND OTHER LOSSES

Every boiler will lose heat by radiation and convection but the heat loss will be reduced if the boiler, its shell or the steam and water drum, are effectively lagged and the lagging is kept in good condition.

The data shown below gives an indication of the magnitude of losses under this heading even from reasonably insulated boilers.

Type of Boiler	Radiation and other Unmeasured losses
Modern Package	2.0
Water Tube (Modern)	1.5
Water Tube (WIF Type)	3.5
Economic Wet Back	3.0
Economic Dry Back	4.0
Sectional	5.0
Cornish, Lancashire, Vertical,	6.0
Lancashire with Economiser	7.0

At first glance, it does not appear as though the operator has any control over the part radiation losses have to play in the attainment of the highest possible thermal efficiency, since these losses are fixed by design.

However, it should be noted that the Radiation losses quoted refer to MCR (Maximum Continuous Rating) and if the boiler plant works on low output the percentage loss is proportionately increased. Also standby boilers which are maintained at pressure incur additional radiation or cooling losses particularly if they are not isolated on the steam and flue gas sides. Seasonal changes in load can be detrimental, e.g. with space heating boilers, burners may work intermittently and substantial cooling losses due to excessive pre- and post-purging and idling can accrue, and in some instances losses up to 10% have been measured due to this!

Load conditions therefore have an effect on the attainment of the highest possible thermal efficiency and it is in this regard that co-operation between Management and Boiler Attendants is most desirable. A boiler cannot work at optimum efficiency when subject to extreme load fluctuations requiring constant alteration to the firing rate and feedwater supply. This is especially true of solid fuel fired boilers, and where much fluctuation occurs, steps should be taken to smooth out the peak demands as far as possible.

GENERAL FUEL EFFICIENCY WITHIN THE BOILER HOUSE

Fuel efficiency can be improved by recovering heat which would otherwise be wasted and using it for preheating the boiler feedwater. Feedwater economisers recover heat from the boiler waste gases before they are discharged, but hot condensate or exhaust steam can also be used to heat the feedwater. Where it is difficult or impracticable to return exhaust steam or condensate to the boiler house, its use within the factory should be examined. A rise of 6°C in boiler feed temperature as a result of heat recovery will save approximately 1% of the fuel. Live steam can be used for preheating boiler feedwater but not for reasons of direct fuel savings. For example, the higher the feedwater temperature, the lower the dissolved oxygen content in the water, and the less is the risk of corrosion; also the total dissolved solids content in the water is reduced due to the reduction in use of sodium sulphite, where this is used to remove dissolved oxygen.

Flash steam escaping from the feedback vent pipe could be recovered by spray quenching and the hot quench water collected in the feed tank.

The rate of heat transfer can be improved in Economic boilers by the use of swirlers or retarders. These are devices which can be inserted inside the tubes and extend the whole length or half length and can be obtained quite cheaply and applied to boilers fired with liquid, gaseous or solid fuel. In the latter case sootblowers may be needed to prevent a build up of loose deposit. Generally, a drop of 18°C in the flue gas temperature results in a 1.0% fuel saving and on a double pass Economic boiler retarders may result in fuel saving of up to 2%. Before using retarders the plant must always be checked to ensure that there is sufficient reserve of draught to overcome the additional resistance of the retarders. It is difficult for a boiler plant to be run at high efficiency for long periods without the help of instruments and strict attention should be paid to the information they provide. If the operator has reason to doubt the accuracy of an instrument, or it becomes sluggish in action, the matter should be reported without delay.

Boiler instrumentation is dealt with in detail in Chapter 12.

OPERATING EFFICIENCY

The maximum attainable thermal efficiency of a boiler depends on its type and design. The following are typical efficiencies.

Type of Boiler	Efficiency (net) %
Vertical Smoke Tube	72
Lancashire	65
Lancashire with Economiser	75
Lancashire with Economiser and Superheater	78
Economic	
(a) Two Pass	75
(b) Three Pass	80
Packaged shell	82
Water Tube	83
Package (low pressure hot water)	80
Sectional	67

HEAT TRANSFER

In a boiler heat is transferred through the metal by conduction, but the heat received by the metal from the fuel bed, flame and hot gases is by convection and radiation. The extent of the contribution of convection and radiation is governed by the design of the heating surfaces and the temperature within the area bounded by the heat transfer surfaces. In the immediate fuel bed or flame area where high temperatures prevail, radiation is the predominant factor.

For example, large furnace tube diameters of relatively long length suited the application of solid fuel firing methods, or of long flame type burners. These surfaces were in sight of the hottest combustion products and of the total heat transferred to the water over 70% took place in the furnace tubes. The effect of standing in front of a radiant type fire is well known, the heat is felt instantaneously.

The effect of heat in areas screened from the radiant fire is not so apparent and a room is slow to warm up at these points.

If a fan type heater is placed in the room, the extremities of the room are heated up rather more quickly by the air which now circulates at a faster rate. Therefore, direct movement of a mass of hot gas transmits heat, and the speed of the air has a profound influence with rate of heating.

In the furnace tube of a Shell boiler, radiant heat is the predominant factor in heat transference. The flue gases are relatively slow moving in the flues of a Lancashire boiler and in order to abstract as much heat as possible, the heating surface area and hence its length is quite great. In order to recover still further quantities of heat a secondary heating surface is added in the shape of an economiser.

The size and space requirement of Lancashire type boilers presented many problems and as the value of velocity of gases in improving the rate of heat transfer was established, boiler designers took advantage of the fact by replacing external flue passages with tube bundles inside the boiler shell.

However, it must be remembered that coal was the principal fuel available in Great Britain at the time when the Economic boiler, as it was known, came into being and the designers had to compromise between heat transfer requirement and the practical one of coping with grit and ash build-up inside the tubes.

It was decades later when clean liquid and gaseous fuels became more abundant that high combustion rates and high convective heat transfer rates in tube nests also became a decisive factor in the design of shell boilers. Now, the so-called modern Package boilers are characterised by being smaller in size, having smaller tube diameters to effect high velocity convection heat transfer compared with boilers of the earlier generation and of like output.

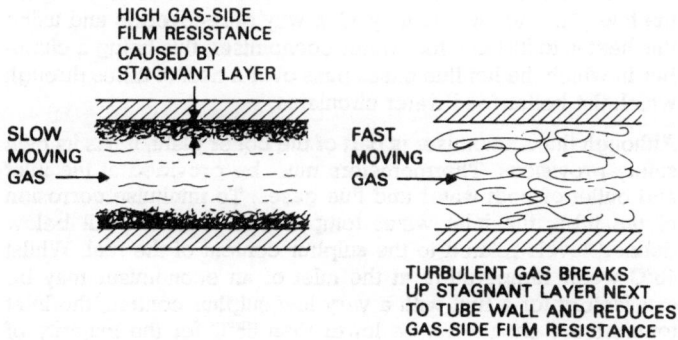

In the illustration above, the relatively slow moving gases have almost a stagnant layer at the metal wall and this layer impedes the heat transfer rate. The fast moving gases in the right hand diagram break down the film and hence its resistance to heat transfer.

The term Package, really refers to the fact that the boiler because of its relatively small size compared with its predecessor can be conveniently sold and delivered complete with its burner and controls.

BOILER HEAT RECOVERY PLANT

Economisers and airheaters should be regarded as direct extensions of the boiler heat transfer surface since surplus heat leaving the boiler itself is used to preheat the boiler feedwater or air used for combustion.

For every 7°C increase in feed-water temperature the boiler fuel consumption is reduced by 1%. The air heater also results in a direct reduction in fuel consumption; there is approximately 1.0% fuel saving for every 20°C rise in air temperature.

FEEDWATER ECONOMISERS

The greatest loss of heat from a boiler is that carried away by the flue gases to the chimney. One way of recovering and using this heat is to install a feedwater economiser, this being a chamber in which the hot flue gases pass over a nest of tubes through which the boiler feed water circulates.

Although the economiser is part of the boiler plant, it has its own safety provisions. Thermometers must be provided at the inlet and outlet of both water and flue gases. To minimise corrosion of the tube, the inlet water temperature must not fall below defined levels related to the sulphur content of the fuel. Whilst 43°C water temperature at the inlet of an economiser may be satisfactory for a fuel with a very low sulphur content, the inlet temperature should not be lower than 55°C for the majority of industrial coal-fired boilers. With fuels of a high sulphur content, the feedwater temperature at the inlet of the economiser should not be less than 77°C. With large industrial water-tube boilers, the inlet temperature to the economiser may need to be as high as 116°C, which may necessitate passing the feed water through an external heat exchanger, heated by means of high pressure steam, before entry into the economiser.

With steam driven feed pumps, exhaust steam can be used for feed pre-heat. Where exhaust steam is not available hot water can be recirculated from the economiser outlet back into the inlet or in the last resort live steam can be injected into the feed tank.

If the water outlet temperature is too high, boiling may occur and cause water hammer which can stop water flowing through the economiser. If the water supply fails, the gases should be by-passed immediately and emergency measures taken to feed water to the boiler.

For safe working, the outlet water temperature of the economiser must be maintained at not less than 22°C below the saturation temperature. If it rises above this limit, part of the flue gases should be by-passed. The following table gives the safe outlet temperatures for steam pressures within the range 60–160lb/in^2.

Boiler Pressure (gauge)		Maximum Temperature at economiser outlet	
lb/in²	bar	°F	°C
60	4.13	267	130
80	5.5	284	140
100	6.9	298	148
120	8.3	310	154
140	9.7	321	160
160	11.0	331	166

Steam tables should be consulted for pressures outside this range and 22°C must be deducted from the saturation temperature. It should be noted that there are some steaming economisers to which these rules do not apply, but these are almost wholly confined to water tube boilers.

SUPERHEATERS

Steam produced from a boiler without a superheater will either be dry saturated, or, more likely, wet. In works where steam is transmitted over long distances, the inevitable heat loss from pipe surfaces causes the steam to become even wetter at the point of use unless a superheater is fitted to the boiler plant. This is a separate bank of tubes placed near the boiler furnace through which steam passes to receive additional heat by convection or radiation. The superheater increases the surface area capable of accepting heat and the production of superheat also slightly increases the thermal efficiency of the boiler. Steam flow must be maintained through the superheater to prevent the tubes burning out and a thermometer should be fitted on the outlet header so that the operator can determine the steam temperature. In a few cases, superheated steam is provided by using a superheater which is fired separately. The maximum steam temperature possible in a shell type boiler is approximately 343°C but this figure can be increased to over 537°C in a water tube boiler.

Superheated steam also improves the performance of turbines and approx. 1% reduction in steam consumption results for every 6°C increase in superheat.

AIR HEATERS

Firing equipment can be designed or adapted to operate with pre-heated combustion air from a plate or tubular air heater placed in the final flue gas stream. This recovers waste heat and increases thermal efficiency. The permissible amount of pre-heat depends on the type of installation. If the furnace gases leave the air heater at too low a temperature, acid may condense out and corrode the metal surfaces. It may be necessary, there-fore, to recirculate air from the heater outlet to the air inlet in order to boost the inlet temperature. In some air heaters glass tubes are used to overcome corrosion with sulphur bearing fuels.

CLEANING HEAT TRANSFER SURFACES

Metal surfaces in the path of the furnace gases need regular cleaning to remove sooty deposits which reduce the amount of heat transfer. Soot blowers are used to remove soot and dust from the tubes of Economic and water tube boilers and the tubes and surfaces of economiser and air heaters. These are high speed steam or compressed air jets, sometimes with a pulsating action to increase the effect. Soot blowers generally have retractable nozzles to prevent damage from overheating when not in use. They are used to remove soft deposits and accumulations of dust in the sole and side flues of Cornish and Lancashire boilers.

Hard or stubborn deposits in water tube boilers can be removed by high velocity water jets known as water lances. These jets are at a much lower temperature than the metal surface at which they are directed and the thermal shock which results breaks off the deposit. Great care must be taken when using them to avoid water coming into contact with the hot refractory brick-work or considerable damage will result.

When using a soot blower on the tubes of an Economic boiler or an air heater, damage to the metal surface by dust and grit erosion will be reduced, if the jet works in the same direction as the normal gas travel. In fact, dust and grit should not be blown around inside a plant but always directed forward to suitable points in the flue system where the accumulation can be removed without shutting down the plant.

If the programme of soot blowing and water lancing is not controlled automatically, e.g. as happens in large boiler plants, steam blowers should preferably be used when the load is low so that the heating surfaces can be at their most efficient when the maximum demand for steam occurs. Water lancing is usually carried out when the boiler load is about 70% or less of its normal rating, though dangerous deposits can be removed by this method when the load is greater.

Chapter 5

BOILER TYPES

A boiler is a vessel used for converting the heat liberated by the combustion of a fuel, into water or steam. It is a pressure vessel, designed to withstand the steam pressure needed in a works, and can be very dangerous if not correctly operated and maintained. An economiser, air heater, or superheater fitted to the boiler will enable even more of the heat liberated from the fuel to be used. Superheaters increase the temperature of steam and are often necessary to render it suitable for use in a turbine or engine.

VERTICAL BOILER

Vertical boilers are used where steam demand is small and ground space is limited. There are two main types:

(i) *Cross Tube*

The simplest form, which is composed of a cylinder containing water with a firebox at the bottom and a flue passing upwards through it to a chimney on top. Cross tubes of large diameter increase the surface through which heat from the hot gases is transmitted to the water. This type of boiler, due to its basic design is thermally ineffecient and has almost disappeared from the industrial scene, although it is occasionally used for mobile cranes.

(ii) *Smoke Tube*

It is desirable to remind boiler attendants of the way heat is transferred, that is, by conduction, convection and radiation.

In this type of boiler, the factors governing effective heat transfer are beginning to become evident in as much as a larger number of smaller diameter tubes pass through the water space, thus increasing the metal surface through which heat is transferred to the water and effectively increasing the flue gas velocity per tube. The tubes usually lie horizontally as shown in Fig. 5. The metal in the flue box above the source of heat is dished thus exposing a greater area of radiant heat receiving surface. All these factors combine to give a relatively good, efficient boiler system.

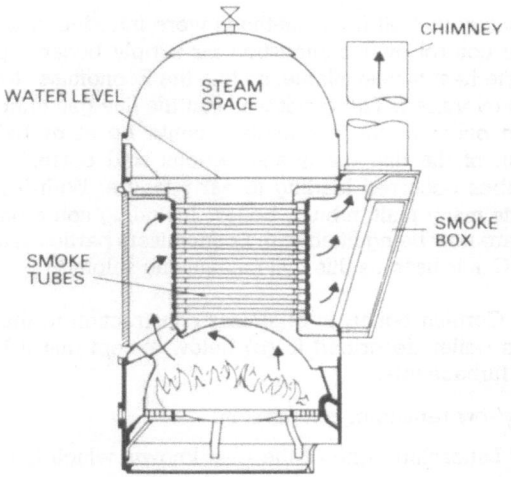

Fig. 5. Vertical smoke tube boiler

Subject to good firing techniques this type of boiler often has a higher thermal efficiency when coal fired than when converted into liquid or gaseous fuel firing methods. The dished surface is ideally located for effective radiation heat transfer from a glowing incandescent fuel bed, although its use is fast declining.

Flue gas temperature leaving the firebox is greater with Gas or Oil and there is inadequate surface area to reduce the flue gas temperature. Many boilers of this type are therefore fitted with retarders to increase heat absorption.

47

HORIZONTAL BOILER

These boilers have an outer shell which contains both the water and the furnace. There are four main types.

The Cornish and particularly the Lancashire Boiler is still used in coal producing regions of the Country. The thermal efficiency of the Lancashire boiler can reach 80% (Net) when fitted with a superheater and economiser.

When heavy fuel oil firing methods were introduced and more effective control over combustion air supply became possible less waste heat was available, so that the Economiser tended to become oversized. The result was that the flue gas temperature near the outlet of the economiser could be at or below the dewpoint of the flue gases and serious acid corrosion of the metal tubes occurred leading to early failure. With increasing fuel costs many multi-tubular boilers including some packaged boilers are now being fitted with Economisers particularly where Natural Gas is used as this fuel contains no sulphur.

(a) The Cornish boiler, is of similar construction to the Lancashire boiler described in (b) below except that it has only one furnace tube.

   Very few remain in commission.

(b) The Lancashire, one of the best known, which is brick-set and has two furnace tubes, each with a grate, mechanical stoker, or burner. The gases from the tubes pass through the boiler shell from front to back below water level, meet and pass through a brickwork downtake, and continue along a brick sole flue under the boiler to the front. The gases then divide and return from front to back through brickwork side flues in which dampers are installed to control the draught. See Fig. 6. Over 75% of the heat is transmitted before the sole flue is reached but a little heat transfer occurs in the sole and side flues. The design is simple, robust, easily cleaned and by reason of size and water content meets rapidly changing steam loads fairly easily. The exit gas temperature is high because of the comparatively inefficient heating surface and an economiser is usually fitted to recover waste heat. Superheaters can also be installed with Lancashire boilers. There is a risk of air

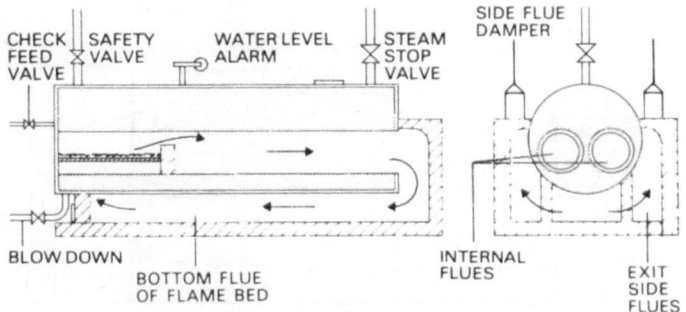

Fig. 6. Lancashire boiler

infiltration through defective brickwork. The Super Lancashire boiler is a modification, without brick setting, in which the gases return through metal tubes used to provide preheated air for the furnaces. Because of the tubes, draught requirements are somewhat higher than in the ordinary Lancashire boiler.

(c) The Economic boiler has a horizontal cylindrical shell and, in modern form, no side or sole flues, although a few of the earlier brickset types still exist. It has one or two furnaces from which hot gases pass through the furnace tubes to the rear of the boiler and enter a steel combustion chamber often lined with firebrick. If this chamber is surrounded by water, the boiler is known as a wetback; if it is outside the boiler shell, the boiler is known as a dryback, the type is illustrated in Fig. 7.

Gases from the combustion chamber return to the front of the boiler through a bank of smoke tubes below water level, enter the smoke box, and pass on to the chimney. This is a two-pass boiler. When the gases pass through a second set of smoke tubes leading to the back-end then to the chimney, it becomes a three-pass boiler. An economic boiler can, therefore be wetback or dryback and of two- or three-pass type. It is essentially a multitubular boiler.

These boilers occupy less space than a Lancashire or Cornish boiler of the same steaming capacity and, because the gases are cooled to a lower temperature before leaving the boiler, they are also more efficient. The small-diameter

49

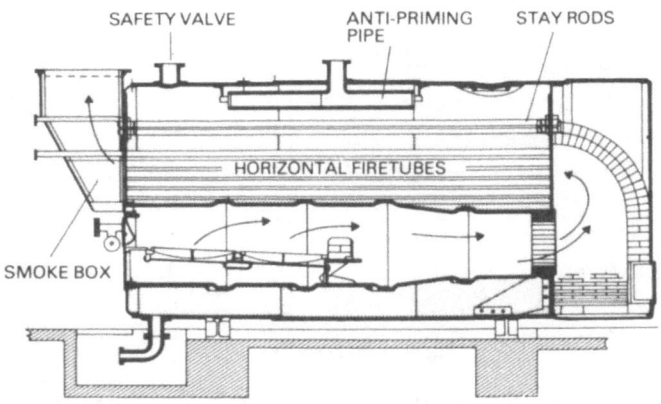

SAFETY VALVE   ANTI-PRIMING   STAY RODS
                PIPE

HORIZONTAL FIRETUBES

SMOKE BOX

Fig. 7. Two-pass dryback economic boiler

smoke tubes create such a resistance to gas flow that it is
often necessary to use artificial or mechanical draught and
although economisers are not usually installed because of
the inherently high efficiency of the boiler a careful assess-
ment of flue gas temperature, operating pressure and load
conditions may well suggest that a feedwater economiser is
economically viable.

(d) The Marine boiler, designed originally for use at sea, is
similar to a wetback Economic, with two or more furnaces,
grates, or burners. It is usually of a large diameter relative
to its length, compared with an Economic boiler.

PACKAGED BOILERS

Packaged boilers are, in effect, highly efficient and very compact forms of an Economic boiler. They may be two- or four-
pass but are usually three-pass, with the final exit for the gases
at the rear. The majority are oil fired and all are mounted on a
chassis or sub-frame on which are installed the pumps for water
and fuel oil, oil heaters, and a control panel for automatic operation. The boiler is assembled and delivered as one unit which
can immediately be connected to the water, steam, fuel, and
electricity supplies, hence the name Packaged. An example is
shown in Fig. 8.

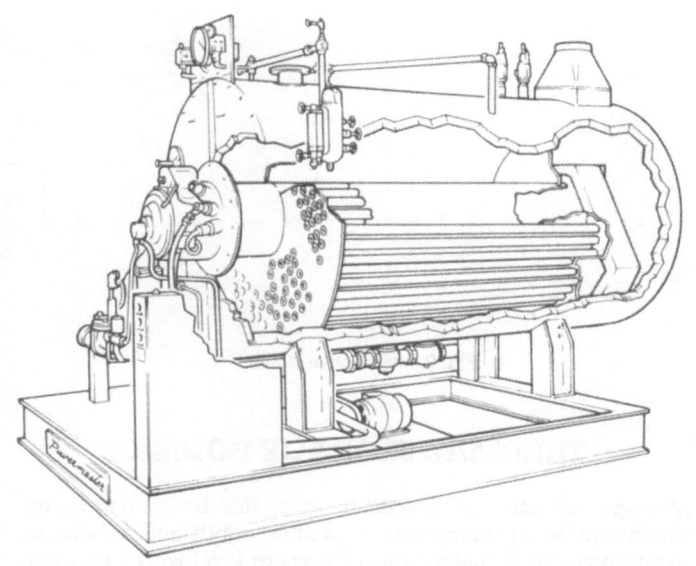

Fig. 8. Packaged boiler

LOCOMOTIVE BOILER

This boiler, like the Vertical type, is used where there is a comparatively small steam demand and is an adaptation, for stationary purposes, of the early pattern of railway locomotive boiler. It has a horizontal shell with narrow smoke tubes running from the firebox at one end to the smoke box and chimney at the other and the gases travel the length of the boiler only once. It is efficient because the firebox above the ashpit is surrounded by water and the narrow smoke tubes expose a considerable surface to the gases. The many tubes have however, to be accommodated below water level, which leaves little space above the water for the steam and can lead to the production of wet steam. See Fig. 9.

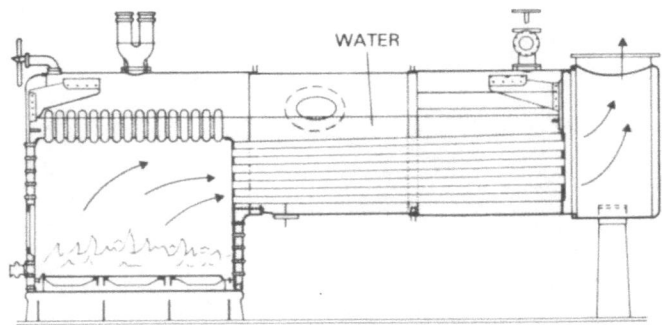

Fig. 9 Locomotive boiler

THERMAL STORAGE BOILER

Although basically an Economic type, this boiler has larger dimensions to accommodate a greater depth and volume of water above the furnace tubes. The water level works between wider limits than are permissible with other shell-type boilers and the firing rate can be kept relatively steady despite peaking loads. At periods of low steam demand, the water level rises and heat is stored. When peaks occur, the level falls and steam output is related to the average heat output plus the stored heat. See Fig. 10.

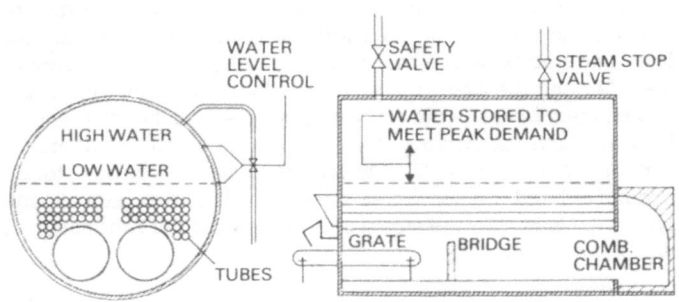

Fig. 10 Thermal storage boiler

WATER-TUBE BOILER

At one time water-tube boilers having steam outputs of 9090 kg/h and operating at quite modest pressures of 17 bar, were common. Such boilers were associated with small electrical power generation, operating a back pressure or pass out condensing unit. A modern shell boiler is quite capable of operating at 23.8 bar, providing steam at 340°C and as much as 27 270 kg/h of steam. Such boilers can provide a power/process requirement suited to modest industrial plant. However, water-tube boilers are still preferred where large steam output, coupled with high steam temperature, are required for power generation or certain processes. This aspect is becoming of less importance since high temperature liquid phase boilers operating at low pressures are now available.

A shell-type boiler needs to be made of heavy construction to withstand higher working pressures and increased ratings. Eventually a limit is reached and a boiler of a different design is required. In these circumstances, a water-tube (W/t) boiler is used, wherein water circulates through tubes connected through drums and/or headers. The tubes are heated by radiation and convection and the furnace is larger than in a shell-type boiler. They can be fired by mechanical stoker, or pulverised fuel, oil, or gas burners, but mechanical stokers are impracticable in the larger boilers because of the grate area required.

The tubes and furnace are enclosed in a brickwork or refractory structure, thermally insulated and, in modern plant, enclosed in an outer metal casing. The radiant water tubes in the furnace, known as water walls, are exposed to the flame or fuelbed.

Other tubes called convection tubes or convection banks, are usually grouped together in nests behind the furnace. The radiant tubes absorb more heat than the convection tubes. Water circulation within the tubes must be vigorous and un-impeded, which means that a high standard of water treatment is essential and blowdown must be carried out to the manufacturer's instructions.

The tubes which carry the hot water and steam upwards are often called risers, while those through which cooler water travels downwards are known as downcomers.

These boilers can be very large, and very complex in construction, and can work at very high pressures. Superheaters may be of the radiant type which receive heat radiated from the burning fuel, or the convective type which take heat from the flue gases at a later stage in their passage through the boiler convection banks. Some boilers have steam re-heaters as well as super-heaters. An economiser, and often an air heater, installed in the flue gas stream beyond the convection tubes are integral parts of the boiler.

The operator of a water-tube boiler must be thoroughly conver-sant with the manufacturer's instructions and the principles in-volved in its safe and efficient operation. A type of water-tube boiler is shown in Fig. 11.

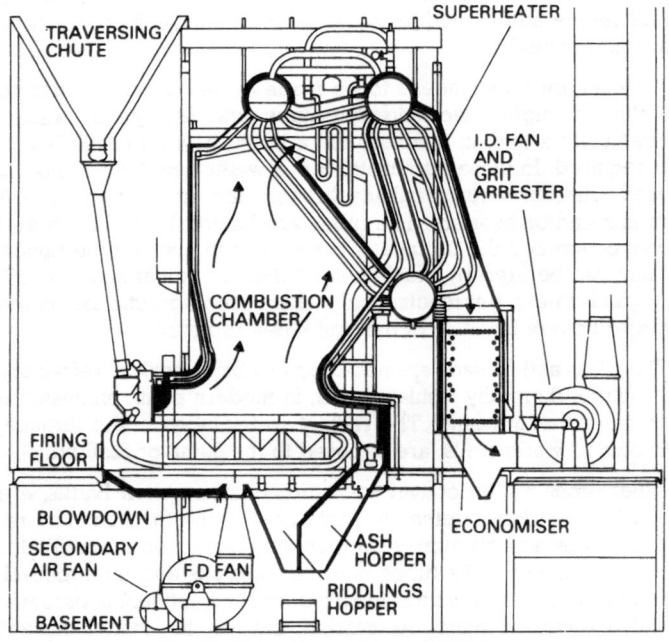

Fig. 11 Three drum water tube boiler

FORCED CIRCULATION BOILER

Some boilers of a basic water-tube design are constructed to operate with a circulation rate very much higher than normal.

Circulation is set up by a pump and the rate varied according to the boiler, sometimes being as high as twenty times the rate of steam demand. See Fig. 12.

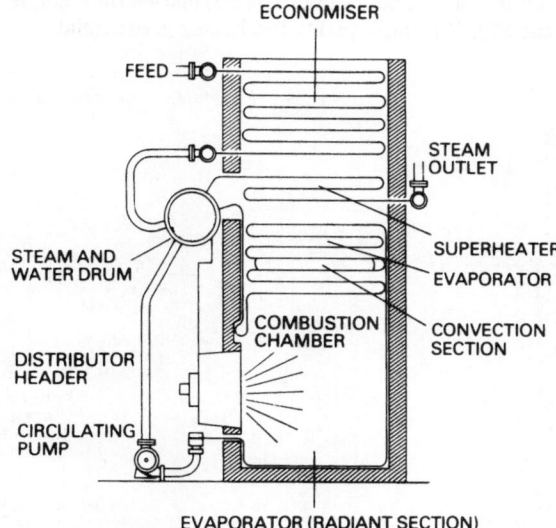

Fig. 12 Forced circulation boiler

STEAM GENERATORS

A steam generator, or coiled tube boiler, is compact and light-weight. It has a rapid steam-making capacity from cold and responds quickly to fluctuating loads.

Water is pumped through the coiled tube, which receives radiant and convected heat and the firing rate is proportional to the load. There are no drums; the feedwater circulates through the coils, partially flashed into steam in a separator, and the remainder recirculated. The size and weight of the boiler are, therefore, considerably reduced.

The combustion chamber is pressurised and heat is released at a higher rate than in a conventional boiler. A combination of forced flow rates on both the gas and water sides of the coiled tube provides high turbulence and velocities for efficient and rapid heat transfer. The fully automatic controls and proportioning system for fuel and air enable the unit to be operated with the minimum of supervision. The boiler is a packaged unit in the sense that it is mounted with its auxiliaries on a single base plate (see Fig. 13). High quality feedwater is essential.

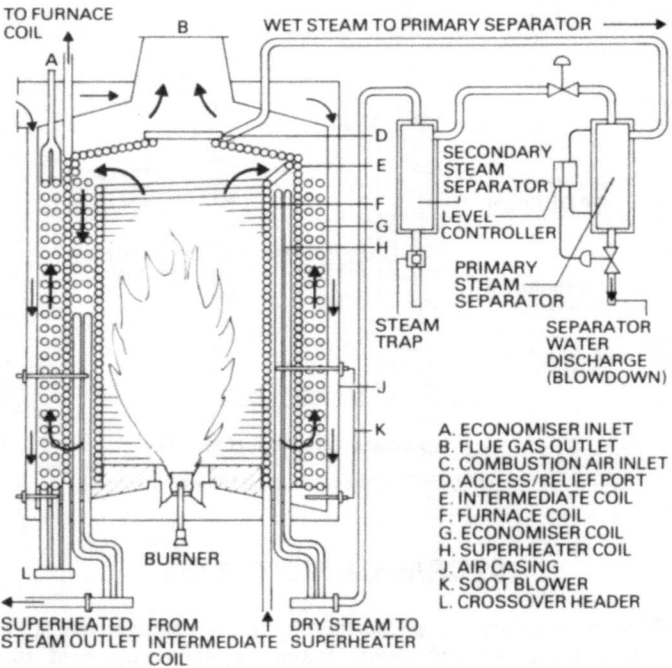

Fig. 13 Steam generator. From Proceedings of 1977 Conference 'Steam at Work' by courtesy of Institute of Mech. Engineers.

SECTIONAL BOILERS FOR CENTRAL HEATING

Most of the boilers described can be used for space heating purposes but central heating is often provided by low pressure hot water. The boilers often used for this purpose are the sectional type, (water temperature usually less than 100°C and are built up from cast iron steel sections which can be added to or reduced in number as required). A typical sectional boiler is shown in Fig. 14.

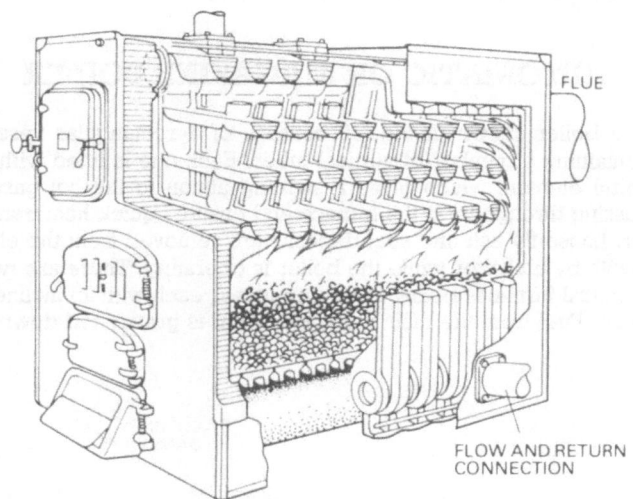

FLUE

FLOW AND RETURN
CONNECTION

Fig. 14 Cast iron sectional hot water boiler

The hollow sections contain the water to be heated and hot water is then carried to the radiators from a flow pipe at the top of the boiler. In all modern systems circulation is by a pump (circulator).

The cooler water returns to a pipe at the bottom. The outer surfaces of the sections provide the heating surface and the majority of them surround the fire where the main heat transfer from the burning fuel to the water takes place. The hot gases then pass to the chimney by way of internal flues in the sections. These flues and the primary heating surface of the firebox must be kept clean by frequent brushing and scraping.

Sectional boilers can be hand or mechanically fired with coke, Welsh steam coal, or anthracite, fitted with oil or gas burners, or fitted with an underfeed stoker to burn coal. A hand fired boiler is designed to operate at full load on one charge of solid fuel every six hours, with sufficient fire left to kindle the next charge. This type of boiler is especially useful where siting is difficult, since it can be transported in sections and assembled at the site. More modern small package boilers are now available for low pressure hot water heating systems.

AUTOMATIC OR MAGAZINE BOILER

The boiler shown in Fig. 15 consists of a rectangular vessel containing a number of vertical tubes. Each one is fitted with a spiral element which gives a swirling action to the hot gases passing through from the furnace and ensures quick heat transfer. Loose fly ash and soot deposits are removed from the elements by a shaker while the boiler is operating. There are two identical furnaces below the water vessel, each with an inclined grate. Fuel from the main storage hopper is gravity-fed down a

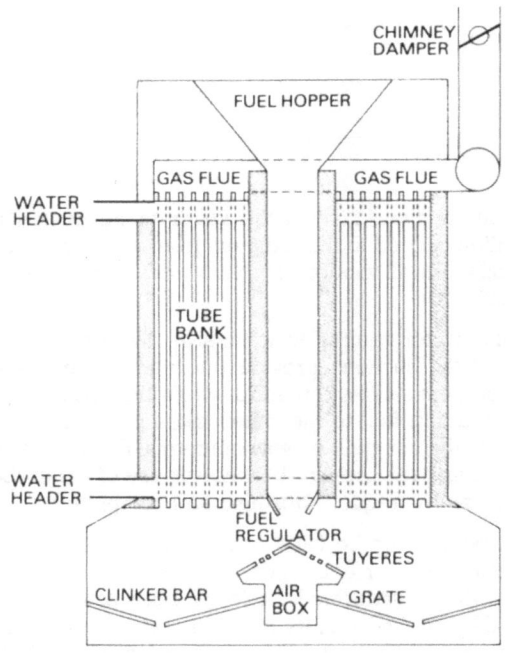

Fig. 15 Automatic hot water boiler

central passage, dividing to give an equal quantity to each furnace. Fuelbed thickness is controlled by a gate pre-set for the type of fuel in use, and a regular quantity of fuel flows to the furnace over a water-cooled tuyere. Air is supplied from a fan in the base of the boiler. Most of it passes sideways into the fuelbed as primary air, though part is used to supply pre-heated secondary air over the fire. The amount of air supplied through the tuyere controls the combustion rate. The fires are cleaned by a ram which discharges ash and clinker over the rear of each grate into the ashpit. This type of boiler can also be used in the smaller central heating installations.

Chapter 6

PRINCIPLES OF SOLID FUEL FIRING

HAND FIRING

Hand firing is not common practice but an understanding of the principles involved help to explain the working of most mechanical stokers. Although of historical interest it could be that in a National Emergency oil burners might have to be removed from converted plant and as an interim measure firebars replaced and hand firing resorted to once again. Furthermore, many people, particularly younger engineers, have never seen a coal-fired boiler; such is the generation gap in coal technology.

When hand firing, it is necessary to supply fuel to the furnace and introduce air through and over the fire so that:

(i) there is sufficient steam to meet demand and maintain a satisfactory pressure without the safety valve blowing;

(ii) smoke and grit emission does not cause an offence;

(iii) the boiler operates at its best practical efficiency.

Solid fuel, of suitable size is added at intervals, the greatest release of volatiles occurs shortly after the fresh fuel is fired. As the volatiles decrease in quantity between firings, the secondary air grid must be adjusted so that, at all times, just sufficient air is supplied to the furnace to burn them off. Too much air will pass through the furnace if there are holes in the fuelbed, a partly-bare grate, or the fire doors are badly fitting. Anthracite and coke need much less secondary air than bituminous coals.

FUELBED THICKNESS

The correct fuelbed thickness for a plant depends on:

(a) the size and kind of fuel;
(b) how much ash it contains;
(c) the boiler load;

and the best depth must be determined by experience. These factors also govern how thick the fuelbed should be midway between firings.

It is difficult to get good results with a fire less than 75mm thick because the primary air passing through the grate will tend to create blow-holes and result in too much excess air. A general guide is to keep the average thickness about 130mm with smalls and 150mm with large coals, such as slacks or singles. The fuelbed thickness may need to be increased if the coal has a high ash content but strongly-caking coals are less troublesome if the bed is kept as thin as possible. The thickness should be 200 to 250mm for coke of the usual sizes and 150 to 200mm for coke breeze. The depth of the bed must be changed according to the load and it should be thicker when a sudden steam demand is likely so as to ensure a reserve of hot fuel on the grate.

METHODS OF FIRING

(i) SPRINKLING
An even layer of fuel is thrown over the grate, with a little less at the back to ensure a bright fire there to burn off the gases and prevent smoke. No part of the grate should be left bare and dead patches must be avoided. The fire is levelled with the rake when the flames are dying down and particular attention must be paid to the back of the grate where there is less fresh fuel. Steam can be raised quicker by the sprinkling method of firing than any other, provided it is used properly, and part of the skill is knowing how to handle the shovel so that the fuel falls on the grate where it is needed. Almost at the end of the throw, the handle should be smartly tilted downwards so that the fuel is 'sprayed' over a wide area. If the shovel is simply jerked away at the end of the throw, the fuel will fall in a heap and make the fire uneven. The knack of doing this comes only with practice.

61

(ii) SIDE FIRING

The fuel is first thrown evenly over the right-hand side of the fire and, when the gases have burned off the fresh coal, the left-hand side is fired in a similar way. This method is strongly recommended because half the fire is always bright enough to burn the volatiles and combustible gases given off by the other half, thus preventing black smoke.

The general rules for both the sprinkling and side firing methods are:

(a) Fire little and often, intervals between firings should not be longer than 10 minutes and just sufficient fuel to maintain the fuelbed thickness should be added.

(b) The fire door should not be kept open longer than necessary.

(c) The grate, including the back and the corners to the left and right of the fire door, must be completely covered with fuel.

(d) The dampers should be adjusted to the load.

(e) The air grids in the fire doors should be opened wider after each firing and the opening reduced when the volatiles have burned off.

(iii) COKING

Fresh fuel is piled on the front part of the grate to a depth of about 250mm and when the volatiles have burned off, the coked fuel is pushed evenly over the rest of the grate with the rake. This method produces very little smoke, provided the right amount of fuel is charged at one time, as the volatiles burn when they pass over the hot fire at the rear. Although this method is not so suitable for high steam output as the sprinkling and side firing methods, it is useful if the operator has other duties because more fuel can be fired at one time and the intervals between firing are longer.

FIRING TOOLS

There are many different types and designs of firing tool but those in general use are:

(i) The shovel, not larger than Size No. 6;

(ii) The poker or pointed bar;

(iii) The rake for levelling the fire;

(iv) The hoe for removing clinker;

(v) The slice bar for separating clinker from the grate;

(vi) The pricker bar for cleaning the spaces between fire bars from below.

Tools should be as light as possible and the use of standard 19mm pipe for the shafts will give sufficient strength and reduce the weight. Tools should only be used when really necessary and care should be taken to avoid knocking the fire about and mixing the burning fuel with ash.

The rake should be passed lightly over the surface of the fire and not driven down on the firebars and pushed along, which will rub soft clinker between the bars and lead to overheating. The fuel and ash will also mix together and it will be difficult to clean the fire. How often the poker is used depends on the type of fuel. With a caking coal, the poker must be used to break up the masses of slow-burning coke which form and occasionally it will be easier to crack the coke crust with the rake. With other types of coal or smokeless fuel, the less the poker is used the better.

GRATE OR FIREBARS

There are three types:

(i) The stationary, from which the ash and clinker have to be removed by hand;

(ii) The self-cleaning, which moves the coal through the furnace and pushes ash and clinker over the back end;

(iii) The rocking-bar, which discharges ash through the bars into the ashpit.

Grates are exposed to considerable heat so are generally made of cast or heat-resisting iron and designed so that the smallest surface is in contact with the fire and the largest with the incoming cool air. To increase the cooled surface, ribs may be cast on the sides of the bars or deep fins provided to distribute the air more evenly through the fire. The air spaces between the firebars depend on the size of the coal. If bituminous coal is burned with natural draught, the spaces should not be less than 6mm wide or more than 19mm wide. With forced draught, they are usually about 3mm wide.

Straight firebars may be cast in sets of two or three, which helps to make the grate more level and prevents bars warping sideways to leave a space through which fuel can drop. These sets are heavier than single bars and are less likely to move and fall into the ashpit when the fire is sliced or cleaned. Straight bars should preferably have a groove running down the centre where ash can accumulate and prevent clinker from sticking. When cleaning the fire, the groove acts as a guide for the end of the poker and makes it easier to remove the clinker.

A self-cleaning grate moves forward through the furnace 50 to 100mm a time carrying the fuelbed with it. Single bars, or groups, then return to the original position without moving the bed. Ash and clinker are pushed over the back of the grate into the ashpit each time the bars move forward.

In the rocking-bar grate, the bars fit across the furnace and rest on pivots. The rocking movement is carried out from the front of the grate and the front and back sections can be rocked independently. This type of grate is suitable for burning high ash coal or coal which tends to clinker since the ash can be removed quickly without the need to keep the fire doors open for long periods during cleaning. With forced draught, trough bars are used to distribute the air through the grate.

REDUCING THE GRATE AREA

The most suitable grate length for a hand-fired boiler is 1.5m and it should not exceed 1.8m, the fire can be better maintained on a short grate. Take, for example, a 9.14m by 2.44m Lancashire boiler, with two grates, each 1.8m long and 1 m wide, which has a total grate area of $3.6m^2$. If 300 kg of coal is burned per hour, the burning rate will be $300 \div 3.66$ or $83.3kg/m^2h$. When hand firing, the most satisfactory results are obtained with a burning rate of around $100kg/m^2h$, although rates of over twice this can be held for short periods. If the burning rate falls to $58kg/m^2h$ or lower, efficiency will be considerably reduced and smoke emission may result from the low furnace temperature.

This can be avoided by shortening the grate sufficiently to raise the burning rate to $100kg/m^2h$. The grate is shortened by covering part of it near the bridge with firebricks, which can easily be removed even when the boiler is in operation. A rocking-bar grate is shortened by fitting dummy bars without air slots.

RECOVERING UNBURNED FUEL

With any type of grate some of the coal will riddle through the bars and the loss is greatest in moving grates or when the air spaces are too wide for the coal in use. Riddlings contain a high proportion of partly-burned fuel and should be re-fired with fresh coal or used for banking.

Clinker and ash also contains some unburned fuel, which varies in quantity according to the ash content and the method of operation. Given a poor fuel and an unskilled operator, a total of one ton in every ten fired can be wasted.

CLINKER PREVENTION

Clinker is caused by ash fusing or melting and the degree to which this occurs depends on the type of ash in the fuel. Coals with a high ash content do not necessarily clinker badly. If the ash has a high melting temperature clinker is unlikely, but with ash of a low melting temperature clinker is practically unavoidable. There are also coals which give little trouble when properly fired but clinker badly when handled carelessly.

Generally, clinker can be prevented by:

(a) using thick fires;
(b) using tools sparingly;
(c) avoiding burning fuel being mixed with ash;
(d) keeping the fire level;
(e) leaving the fire undisturbed for as long as possible and then thoroughly cleaning out.

A wisp of steam under the grate will reduce clinker formation but this method should only be used when all others have failed or steam will be wasted. Very fine jet water sprays can also be used successfully.

CLEANING FIRES

Before cleaning, the water level in the boiler should be increased and the fire allowed to burn down until there is just sufficient hot fuel to start up again quickly. When the fire is in the right condition, the dampers must be partially closed. If the fire is thin, it may be necessary to add fresh coal to the side which will be cleaned last. The tools to use are the slice bar, the hoe, and the rake.

There are two methods of cleaning:

(1) SIDE METHOD
   (i) The fuel is pushed from the left-hand to the right-hand side of the grate leaving the ash and clinker.
   (ii) The ash and clinker are then loosened with the slice bar and scraped out with the hoe.
   (iii) The fuel is pushed back from the right-hand to the left-hand side of the grate and fresh fuel is fired on to it. Additional fuel is added to the right-hand side if it looks like burning out.
   (iv) When the fire is burning well, the ash and clinker on the right-hand side are removed.
   (v) The fire is then spread evenly over the grate and built-up with fresh fuel.

(2) FRONT AND BACK METHOD
   (i) The burning fuel is pushed against the bridge wall with the rake or hoe.
   (ii) The clinker in the front is loosened with the slice bar and pulled out of the furnace with the rake.
   (iii) The fire is spread evenly over the grate and built up with fresh coal.

With the front and back method, some clinker and ash are always left at the back, which may put part of the grate out of action. Clinker may also stick tightly to the bridge wall and be difficult to remove. This method of cleaning may be necessary when there is a high steam demand but it should always be followed by a thorough cleaning with the side method when the load drops at the end of the day.

Cleaning preferably should be fitted in with periods of low load. If, however, the fire urgently needs attention when the load is high, the slice bar can be run underneath it, slightly twisting the handle while the bar moves sideways. This will loosen, and possibly break up the clinker though care must be taken not to lift any of it into the upper part of the fire.

When the ash and clinker are removed from the fire, the live fuel must not be pulled out with it. The slice bar can be run along the grate to break up the clinker and detach it from the bars but must not be used to lift clinker through the fire or fuel will fall through the firebar spaces and clinker formation will be aggravated.

If there is more than one furnace and it is necessary to clean the fires while the boiler is under load, the following routine may be adopted:

(a) The left-hand side of all the furnaces should be cleaned in turn.
(b) When these are burning freely, the same procedure must be adopted for each right-hand side of the furnaces.

BANKING

When the fire is banked, fuel is left on the grate and allowed to burn away very slowly, using just enough to compensate for natural cooling or to produce any small amount of steam needed during the banking period. The bulk of the fuel on the grate should be red hot before starting to bank. The grate is then cleared and the live fire is pushed against the bridgewall. Clinker on the rest of the grate is loosened and left lying with the ash to prevent air passing through the front and middle of the grate. Fresh fuel is put on top of the fire at the back. With a self-cleaning grate, the fire is pushed to the back of the grate and the self-cleaning action stopped. Sufficient fuel is then added, the air grids in the fire doors are closed and the dampers are left sufficiently open to prevent fumes escaping into the boiler house. It is usually possible to close the dampers a little further about 30 minutes later. The alternative to 'banking' is 'drawing' the fire which literally means cleaning out the grate completely at night and re-kindling the following morning.

Fortunately most boilers have automatic feedwater regulation. Where the system is manual however, the water in a banked boiler must be left higher than the normal working level and the feed valve, blowdown cock, and crown valve must be tightly closed to prevent the loss of water.

Unburned gases accumulate and become a potential danger in a banked boiler. Before starting up again, it is essential to purge the system by raising the chimney damper and opening the air grids before the fire doors are opened. When the system is clear, the furnaces are cleaned and the fuelbed built up to the required thickness.

Boilers are sometimes banked overnight to ensure a quick start in the morning, but provided the boiler setting and insulation are in good order and the dampers tightly closed, pressure can be re-gained nearly as quickly and fuel is saved, if the fires are 'drawn'.

Chapter 7

MECHANICAL FIRING

The majority of industrial boilers burning solid fuel are fired mechanically rather than by hand for the following reasons:

(a) fuel is more evenly distributed over the grate;

(b) there is less risk of smoke emission;

(c) furnace doors do not need to be opened at frequent intervals either for firing or for removal of ash and clinker;

(d) Steam output is increased;

(e) operating conditions are steady for a given load.

(f) the physical effort in firing boilers by hand is considerable.

TYPES OF MECHANICAL STOKER

1. SPRINKLER

Sprinkler stokers are rarely used now since their popularity has declined because of a tendency to cause grit and smoke emission. As the name implies the stoker imitates the sprinkling method of hand firing. Coal is fed in measured pre-set quantities from a hopper to a box at the front of the furnace tube and the box contains a spring-loaded shovel or flipper which flings the coal lightly and evenly over the grate area. In this stoker, the fuel throw is varied, one stroke putting the fuel to the back of the grate, the next to the middle, the third to the front, and so on. Another type has a 'beater' shaft with blades that rotate at high speed. Coal is metered into the beater box and propelled onto the grate (Fig. 16). The grate can be stationary but is generally of the self-cleaning type.

The sprinkler stoker operates with natural or forced draught,

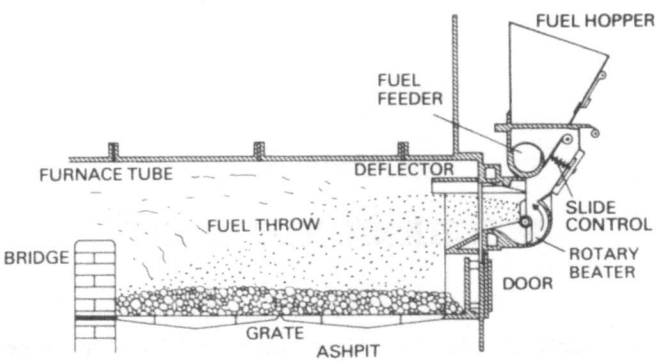

Fig. 16 Rotary type sprinkler stoker

responds quickly to changes in steam demand, and has an easily controlled rate of fuel feed and distribution but, when the boiler is shut down quickly and later re-started, smoke and grit emission are difficult to avoid.

The fuel rate is altered by changing the stoker speed or the length of feed-ram travel. Fuel distribution can be varied by adjusting the height of the cams and the tension of the spring operating the shovel or altering the speed of the rotary distributor, or the angle of the deflector plate. Fuel distribution should be checked occasionally by running the stoker when the boiler is idle and adjustments made to ensure that the grate is evenly covered without coal being thrown over the back. Faulty distribution can be caused by bent or worn shovels, worn tappets or cams, or badly-adjusted tappets or springs.

The best results are obtained with graded fuels, such as doubles, singles or peas, which give a more even fuelbed. Smalls are not recommended since they tend to give a much thicker bed in the centre and can cause excessive carry-over of grit. The fire should be approx. 70 mm thick when burning peas and 100 mm when burning singles or doubles. A thin fire responds more quickly to load variations but requires more care and attention to keep it in good condition. Thinner and more even fires are obtained with a rotary-type sprinkler.

Burning rates up to 195 kg/m²h are possible but grit emission is serious above 145 kg/m²h. If the load is well below the boiler

70

capacity, the grate area should be reduced and the stoker throw adjusted.

Draught should be just sufficient to keep a light haze at the top of the chimney and less draught is needed for a rotary-type stoker. When using natural or induced draught, smoke emission can be prevented or reduced by adjusting the secondary air grids on the furnace doors. If forced draught (FD) is used, some of the air from the FD supply is used as secondary air.

The fires on stationary grates are cleaned in the same way as when hand firing but the stoker must be stopped during the cleaning period. Clinker on the firebars should be eased with a flat slice bar and removed with a rake, not lifted and left in the fuelbed. Even with self-cleaning grates, clinker will probably have to be removed by hand.

The selected grate speed will depend more on the amount of ash to be moved per hour than on the burning rate. If the speed is too high, quantities of riddlings will fall between the firebars and unburned fuel will also be carried over the end of the grate. It is generally necessary to cool the firebars by fine water sprays, or the use of a little steam.

2. COKING

This stoker imitates the coking method of hand firing. There are several different types but in each one the fuel is fed from an overhead hopper to the front of a reciprocating piston or ram. In the type illustrated in Fig. 17, the ram pushes the fuel on to a coking plate, where it ignites and finally is moved towards the ash discharge end by the reciprocating action of the firebars. Volatile matter released early in the travel of the fresh fuel is ignited and burned without smoke as it passes over the incandescent fuel at the back of the grate. The length and frequency of the ram stroke can be altered to vary the amount of fuel fed to the furnace. Ashes are removed by use of a long shovel by first opening the hinged sealing door at the rear of the grate. It is necessary to keep this door in good repair to reduce air in leakage.

The grate of a coking stoker is formed of cast iron or heat-resisting steel firebars which are driven backwards and forwards by a system of cams. First, all the firebars move forward, taking the fuelbed with them. Second, individual or groups of

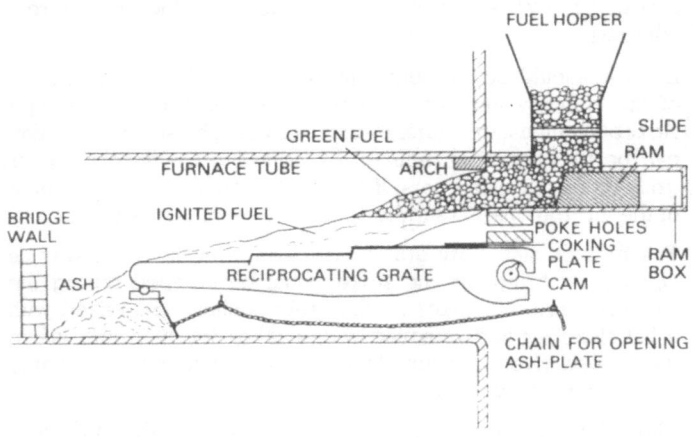

Fig. 17 Ram type coking stoker

bars return to their original position, leaving the fuelbed forward. Each time bars move backwards, ash and clinker are discharged from the end of the grate.

A wide range of coal, from smalls to singles, can be used on these stokers. The ash content of the coal can be as high as 12% but, unless special firebars are used, should not be less than 6% or overheating and burning out of the bars will occur. Coals with pronounced caking tendencies are not suitable for a coking stoker, although weakly-caking coals tend to reduce riddling losses. The slice and poker should be used sparingly.

The main advantages of a coking stoker are:

   (1) it can burn a wide range of coals, except those with pronounced caking qualities;

   (2) it needs little attention apart from infrequent slicing of the fuelbed;

   (3) if care is taken with the air supply and the fuelbed left undisturbed, smoke is unlikely;

   (4) there is little risk of grit emission, even when the coal contains a large amount of fines.

The fire should be 300 to 350 mm thick at the front of the grate, tapering off towards the back, with the whole grate covered. In older types of stoker, with a high narrow feeding ram (high ram) there is a tendency for partially-burned coal to pile up in the middle of the grate leaving the sides thinly covered. This can be avoided by fitting a V-shaped projection in the centre of the top coking plate so that the coal is thrust towards the sides. The trouble is not likely to occur with wide and low rams, as fitted to modern stokers.

An over-thin fire which allows too much air to pass through the bed can be cured by increasing the coal feed or reducing the draught. If the fuelbed is too thick at the back, partly-burned coal can be carried into the ashpit. This can be prevented by reducing the coal feed or increasing the draught according to the load on the boiler.

Coking stokers are usually designed for a maximum burning rate of 170 kg/m^2 but rates of 244 kg/m^{2}h or more have been obtained with shortened grates when using certain specific types of coal. Control is difficult if the burning rate is less than 58 kg/m^{2}h. If the rate is consistently low, the stoker manufacturer should be asked to replace the bars so that the grate area can be reduced and the burning rate increased. If the load is reduced for a short time, or for only part of each day, an air deflector plate can be fitted under the grate. This is less effecttive than shortening the grate but better practice than using the full grate on reduced load.

Induced draught is generally used but the thick fuelbed and small area of air passages through the grate make draught requirements high. The amount of draught over the fire will vary with the burning rate, but the following table is a general guide.

Burning Rate per hour		Draught	
kg/m^2	lb/ft^2	mbar	in. wg
98	20	0.6	0.25
122	25	1.0	0.4
147	30	1.3	0.5
171	35	1.7	0.7
195	40	3.0	1.2

The firebars can become red hot if there is insufficient draught or if clinker prevents the air from passing through and cooling them.

73

The coking stoker, with its thick fire, is slow to respond to changes in steam demand and steam users should give the operator advance notice if the load is likely to be varied greatly.

When necessary, the fire should be sliced by passing the poker through it at grate level without unduly disturbing the fuelbed. Some coking stokers are provided with ports for this purpose.

3. CHAINGRATE

This stoker has no counterpart in hand firing. The grate is an endless chain fed with fuel from an elevated hopper via a guillotine door. The chain carries the fuel under an ignition arch, where it starts to burn, and ash and clinker are discharged over the back end into the ashpit (see Fig. 18). These stokers vary in detail and the manufacturer's instructions should be carefully followed. They readily adapt to full automatic control and need little attention. In some cases, provision is made for mechanical ash and clinker removal. There are also 'mini' chaingrate stokers designed for use with sectional and central heating boilers and some small steam boilers.

The guillotine is raised or lowered to vary the thickness of the fuelbed and the chain speed altered to suit the boiler load or the type of fuel in use. Some of the stokers are fitted with compartmented forced draught. This enables the air supplied

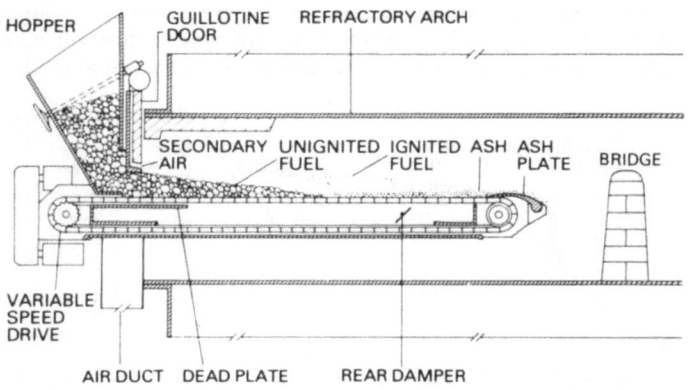

Fig. 18 Chain grate stoker

from beneath the grate to be varied along its length, which keeps the fire even and ensures that ash and clinker are satisfactorily burned out on discharge. The air at the back of the grate can also be reduced to prevent the fuel burning out too soon and leaving the back end bare.

These stokers usually work best under balanced draught conditions. They are fitted with a forced draught fan to raise the pressure of air under the grate sufficiently for it to pass through against the resistance of the chain links and the fuel. Secondary air, damper-controlled, can be admitted over the fire through separate ducting and passes into the furnace through ports in the front of the furnace door or ignition arch; but in practice is often not required with this stoker. Draught requirements are less than those required for a coking stoker.

A wide variety of fuels can be burned but the most suitable are smalls or small graded coal. Coke breeze can also be burned but advice should be sought on this. Ash content should be not less than 8% to ensure that a sufficient thickness of ash rests on the chain links to prevent overheating. Fuels sometimes need moisture conditioning and should be just moist enough to 'ball' when squeezed in the hand. It is usual to have the fuelbed about 100–130mm thick and level across the width of the grate at the guillotine. An uneven fuelbed can result if the guillotine door is out of alignment. Coal distribution over the grate will be affected if the fuel in the hopper is too low and the fire will run off the end of the grate if the hopper runs empty. Generally, the stoker should be operated with as little interference as possible but, with caking coals, it may be necessary to run the slice along the sides of the grate to prevent fuel sticking to the furnace tubes and causing an uneven fuelbed.

The grate speed can be varied over wide limits up to approximately 12.2 m/h and should be controlled according to the boiler load, making sure that it is not so high that live fuel is dumped over the end of the grate into the ashpit. The burning rate varies from as low as 73 kg/m²h to well over 170 kg/m²h.

A boiler fitted with a chaingrate stoker can be banked. The level in the fuel hopper should be allowed to fall during the 30 minutes beforehand, the guillotine door then being raised about 50 mm above normal to empty the hopper on the grate, and the grate and forced draught fan stopped to leave a heavy bank of fuel just inside the fire door. The door is then lowered as far as possible and gaps underneath sealed with fine ash. The flue dampers are left just sufficiently open to prevent emission of smoke from the boiler front. The ashpit is cleaned out and the doors closed. Care is essential to ensure that the links are not overheated.

When the boiler is re-started, the bank of fuel is broken up and raked back to mix with the fresh fuel. The hopper is filled with fuel, the guillotine raised to the normal position, and the forced draught fan started. The grate is then started at lowest speed which is increased as the ignition arch heats up.

4. AUTOMATIC CONTROLS

Both coking and chaingrate stokers fitted to shell boilers can incorporate automatic combustion control equipment, where the fuel/air ratio, once set for a given fuel, can be maintained. The operation of the stoker can also be re-adjusted in relation to load changes, usually by reference to the maintenance of steam pressure. The former type stoker is usually controlled by the Unitherm system, whilst the latter frequently incorporates the Handimatic system, to mention but two proprietary systems.

It is difficult to control a mechanically fired boiler of the moving grate type to the same extent as an oil or gas fired unit. For example, if a mechanically fired boiler of the moving grate type is used in a school and a time clock shuts the system down at 4.00 pm, then difficulties could be experienced at start time in the morning. An unattended boiler would be left with a body of coal on the grate, since at shutdown there would be little cooling air under the grate, clinker formation would result. If the clinker formed is of the biscuit type which effectively seals off the air spaces, there would be little combustion air available during start-up and steam pressure or hot water temperatures would not be attained. Further, the moving grate would carry over excess unburnt fuel and cause damage to ashpit seals. Other allied problems could also occur, dependent on the type of coal, layout of flue system etc.

5. UNDERFEED

Unlike the stokers described above, the underfeed system can be made almost as automatic as oil and gas fired boilers.

As with the chaingrate type, the underfeed stoker has no counterpart in hand firing. Fuel is delivered from a hopper or outside bunker by a conveyor screw or ram to a retort. As the fuel passes up through the retort area, combustion air is supplied under pressure to jets or tuyeres placed near the top or upper sides of the retort, at which level combustion takes place. Volatile matter from the fresh coal rises through the incandescent coal on top of the retort, reaching ignition temperature as it meets the air supply. This ensures smoke-free operation. See Fig. 19.

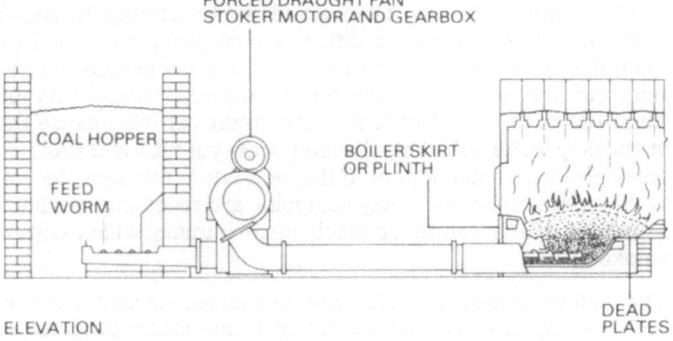

FORCED DRAUGHT FAN
STOKER MOTOR AND GEARBOX

COAL HOPPER

FEED WORM

BOILER SKIRT OR PLINTH

ELEVATION

DEAD PLATES

Fig. 19 Hopper model underfeed stoker. By courtesy of NCB/College of Fuel Tech. (London)

The circular 'pot' type retort is generally used for simple installations. The 'grate' type shaped like a rectagonal box, can be installed as a duplex or multiple unit with two or more retorts when a bigger input or heat is required. In both types, the ash and clinker spills over the retort on to the side plates and is raked out.

Combustion air is delivered under pressure to the tuyeres by fan and controlled by damper. Over-fire draught requirements are only about 0.12 mbar negative.

The best fuels are free-burning graded coals with a low proportion of fines and moisture, since both these characteristics increase the tendency of the fuel to arch-up in the hopper or block the screw conveyor. Peas, beans, singles and, in the larger stokers, doubles, may be used. Coking coals are unsuitable because of a tendency for coked fuel to climb up through the

77

centre of the retort, producing a 'coke tree' or 'cauliflower' which has to be broken down manually. This condition disturbs the even burning of the fuel in the retort and often causes smoke. Coal containing an excess of fines will segregate in the bottom of the retort and eventually come up through the end of it furthest away from the feed as a dense mass which excludes air and causes that part of the retort to go dead.

Underfeed stokers can be fitted to sectional and central heating, vertical steam, and other horizontal shell boilers.

The rate of fuel feed, which should ensure that the fire just mushrooms over the top of the retort is set by altering the speed of the motor or selecting a different gear ratio. With too little feed, the fire recedes inside the retort, the uncovered tuyeres burn out and, eventually, the end of the feed screw may be burnt off. A low level of fuel in the retort can also cause the volatiles to leak back along the feed conveyor tube and produce smoking-back in the hopper. If the feed rate is too high, the fuel in the centre of the retort may not ignite, and result in a condition known as black centre or black heart burning which causes smoke.

The underfeed stoker usually operates under automatic control and is stopped or started according to the steam pressure or the water temperature. In some designs, the single motor driving the fuel feed and forced draught fan is associated with a device which automatically varies the air damper setting to suit the amount of fuel being burned and prevents fuel burning out during idle periods. With automatic on/off control, the selected fuel feed should be the minimum required to run the stoker for the maximum time between stops, thus preventing ingress of cool air through the furnace.

The stoker must be switched off before the fires are cleaned and the intervals at which ash and clinker are removed can only be determined by practice and experience.

When the stoker is started up, coal must be run through until the tip of the worm is just covered. The rest of the retort is then filled up with coke or smokeless fuel to tuyere level and paper and wood are added. The fire is lit and while the feed is stationary, a little forced draught is used. Coke is added to the ignited wood and the fire built up until a good red body of coke covers the tuyeres. At this point, the stoker should be started to feed coal through the worm. If this procedure is followed carefully, virtually smokeless lighting-up can be guaranteed. To shut down the stoker, the air supply should be reduced progressively so that the fire cools gradually and the tuyeres are not damaged. The coal feed and air supply are then stopped and the chimney damper closed to prevent air being drawn in to keep the fire alight.

In an emergency, it may be necessary to fire the furnace by hand. Then, the base of the retort and the feed tube should first be protected from damage by placing firebricks in the retort up to the level of the tuyeres. The coal feed should be set in the neutral position and the motor switched on with only the fan running. If the motor breaks down, the best available draught is provided by opening the air control to its fullest extent and removing the cleaning doors on the air duct leading to the retort.

The coal feed, whether screw or ram, can become blocked by large pieces of coal or tramp iron and the fuel hopper should be protected by a metal grid or wire screen. A sheer pin is often fitted to prevent damage when the feed mechanism jams. If it breaks, the cause should be discovered and remedied before a replacement is inserted, the new pin must be the same size and strength as that originally supplied by the stoker manufacturer.

Where a stoker is used with induced draught, the controls should be interlocked electrically with the fan drive so that the draught starts slightly ahead of the fuel feed and stays on for a few seconds after it stops, thus avoiding a blowback, and smoke formation.

Underfeed stokers, especially on central heating boilers, are often fitted with a 'kindling' clock control, that brings the feed motor, and forced draught fan, and any induced draught fan, into operation for a short period every hour. This keeps the fire alight in the retort and the boiler and furnace can then be brought into full operation by a clock control as and when required. The optimiser control system can also be applied whereby the starting time can be varied in accordance with outside temperature. The time the kindling control is in use depends on the type of installation and the fuel, but should be as short as possible to prevent ash and clinker being forced out of the top of the retort while kindling. This must be removed quickly or the entire side plate area will become choked and the ash and clinker will spill back into the retort.

Coke is not generally used on an underfeed stoker because it causes abrasion of the feeder gear and tube.

6. MAINTENANCE OF MECHANICAL STOKERS

A mechanical stoker works under the most stringent conditions of wear from the fuel and extreme heat from the fire. To give satisfactory service, it must be thoroughly cleaned at regular intervals and the moving parts kept as free as possible from coal or coke dust. Preventive maintenance is considerably more effective than breakdown maintenance and cheaper in the long run.

The oil level in the gear boxes must be maintained and the oil changed every two months. Oil or grease lubricators must be filled each shift and grease nipples given periodic attention. The oil and grease should be of a type recommended by the stoker manufacturers and stored in sealed containers to prevent contamination by dust or grit. The stoker should be inspected at regular intervals to ensure that it is working as the manufacturer intended. Adjustments should be made and defective parts replaced as soon as possible.

Although the effective control of a mechanical stoker needs as much skill as firing a boiler by hand, the equipment relieves the operator of a considerable amount of manual labour.

7. VEKOS AUTOMATIC AUXILIARY FURNACE

The Vekos furnace, or stoker, as fitted to a Lancashire, Economic, or Cornish boiler, consists of a cylindrical, water cooled combustion chamber connected to the boiler furnace tube or tubes by a refractory-lined steel duct, with a close fitting door at the end of the furnaces for cleaning purposes. Water for cooling the combustion chamber is taken from the bottom of the boiler to the front end of the water jacket and returned from the top rear end of the jacket to the boiler at a point just below normal water level. By this means, radiation losses are reduced and water circulation within the boiler improved. The Vekos stoker has been successfully built into packaged boilers, and there are many successful installations. (See Fig. 20.)

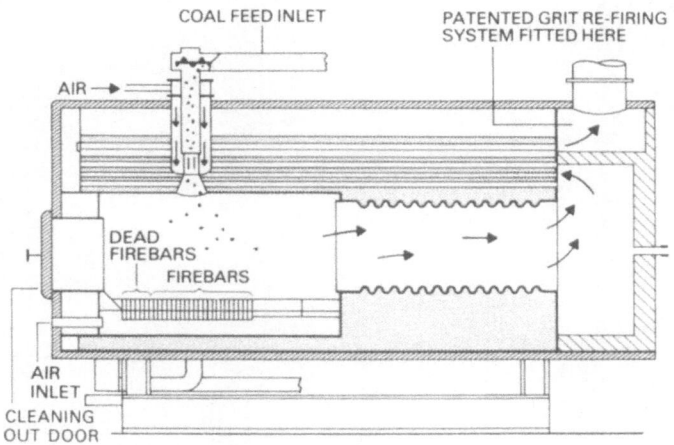

Fig. 20 Vekos. By courtesy of NCB/College of Fuel Tech.(London)

The grate is fitted at the bottom of the combustion chamber and has no moving parts. Coal is fed at a variable rate into a whirling chamber at the top of the furnace by a worm conveyor located immediately above it. The coal is swirled about by air blown through slots arranged tangentially in the walls. This ensures good distribution over the grate surface. A single fan supplies combustion air, as well as air for the whirling chamber, to an air box at a pressure of between 1.0 mbar and 3.0 mbar. The box, located below the furnace, has control dampers. Primary air passes from it through slots in the gratebars and secondary air passes through slots in a dead plate situated between the front end of the grate and the bottom of the furnace door. Tertiary air can be made available around the joint between the end of the furnace and the boiler.

This stoker needs little attention. Ash and clinker are usually removed only once in every 8 hours and, as the fire does not need poking, combustion is free from smoke and soot. Power requirements are low and there are no moving parts in the furnace. Bituminous smalls can be fired satisfactorily and high average CO_2 content can be maintained in the flue gases.

Chapter 8

PULVERISED FUEL

Pulverised fuel (PF) is coal, or sometimes hard pitch obtained from the distillation of coal tars, which has been ground into small particles. A good grade of PF should have particles of not greater than 1/200" (0.13 mm) and between 75% and 85% should pass through a 200 BS sieve. The finely-ground fuel is burned at a nozzle or jet in a similar way to gas or liquid fuel.

Puliverised fuel is used mainly in power generating stations or cement kilns, but is rarely used in the small steam raising plants due to the problems associated with containing the flame within the fire tube length, but more particularly due to the large capital requirement on plant to remove the fine grit and dust particles from flue gases and the cost of pulverising equipment.

It has several advantages

    (i) cheaper and lower grade fuels can be burned as PF;
   (ii) the burners readily respond to load changes;
  (iii) banking losses are practically eliminated;
   (iv) automatic control can be used;
    (v) the position of the flame and therefore, the hot or high temperature zone in the furnace, can be altered;
   (vi) very high air pre-heat temperatures can be used;
 (vii) steam output is higher than when burning solid fuel on grates.

A boiler using PF can produce over 450 tonnes/h* of steam and the largest steam-raising units are fired with this fuel. There are however, some disadvantages. The capital cost of pulverising plant is considerable and the plant occupies a lot of ground

* 1 tonne = 1000 kg = 2200 lb.

space. The coal has to be dry to obtain the right degree of fineness. Grit emission is heavy and the installation of complex grit arrestors is essential on PF plant.

DRYING THE FUEL

If the coal has surface moisture, or is stored in the open, it must be dried before pulverising, the drying can be done inside the pulverising mill by passing pre-heated air through the coal before the air is used for combustion purposes. The warm air can be supplied by an air heater placed at the back of the boiler behind the economiser and the normal combusion air heaters, or it can come from a separately fired pre-heater. If the uncrushed coal is very wet, a separate drier can be used before the coal is taken to the grinding mills.

CONVEYORS AND STORAGE

Coal is conveyed to the mills or pre-driers by conventional handling plant, conveyor belt, or elevator but, when ground, a closed system is necessary to prevent considerable losses. In the Unit, or direct-fired, system the pulveriser is situated as close as possible to the boiler and the fuel feed rate into the mill is controlled at the rate that the fuel is burned. Some combustion air is used to convey the PF from the pulveriser to the inlet of the fan feeding the burners and clean air is used if the mill is pressurised to reduce the need for fan maintenance. The fan speed, or the degree to which the fan damper is opened, controls the velocity of the air stream bearing the PF through the ducts connecting the pulveriser to the boiler and must be carefully controlled if particles of the right size are to be air borne.

In the second system, dried fuel passes through the pulverisers and is airborne to a central cyclone-type separator where the PF is extracted and collected into a bunker or hopper which takes up variations in supply and demand. The fuel is then picked up by an enclosed worm or screw feeder and delivered at the correct rate into the primary air stream to the burner. See Fig. 21.

Pulverised fuel can be purchased for installations without pulverising plant. It is delivered into a storage hopper or bunker

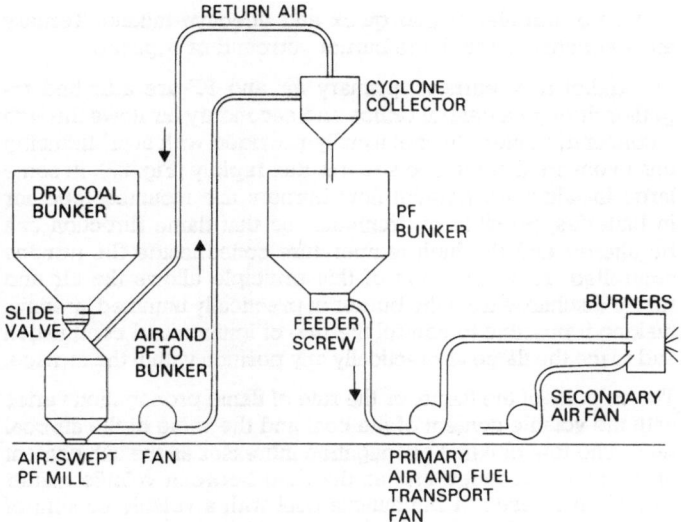

RETURN AIR

CYCLONE
COLLECTOR

DRY COAL
BUNKER

PF
BUNKER

BURNERS

SLIDE
VALVE

AIR AND
PF TO
BUNKER

FEEDER
SCREW

SECONDARY
AIR FAN

AIR-SWEPT   FAN
PF MILL

PRIMARY
AIR AND FUEL
TRANSPORT
FAN

Fig. 21 Pulverised fuel system

and thereafter handled in the manner already described. In a multiple installation of boilers and/or burners, a ring main system can be used to distribute the PF from the hopper. The fuel is air borne in the ring main and any unburned surplus is separated out in cyclones at the hopper.

PULVERISED FUEL BURNERS

A PF burner should give rapid and intimate mixing of the fuel with the combustion air and must produce a flame shape suitable for the furnace. A 'turbulent' flow burner produces a short flame and a 'parallel' flow burner a longer one.

Since the fuel is transported in air some combustion air is delivered into the burner with the PF, secondary air entering the furnace through an air register surrounding the burner similar to that used with liquid fuel burners. If a short intense flame is needed from a turbulent flow burner, secondary air is admitted round the jet through which the primary air and the fuel are passed, the two streams meeting at right angles with a high degree of turbulence and quick and effective mixing. Tertiary air is admitted through the burner surround or register.

In parallel flow burners, primary air and PF are admitted together through a central orifice and secondary air flows through a concentric outer channel usually provided with swirl-inducing fins to ensure that the two streams mix rapidly (Fig. 22). In some large installations, parallel flow burners are mounted together in batteries, possibly on trunnions, so that flame direction can be altered and the high temperature zones inside the furnace controlled. A modification of this principle allows the air and fuel to discharge from the burner in practically unmixed streams, making it possible to control the rate of ignition and combustion and place the flame at practically any position within the furnace.

The velocity of the flame, or the rate of flame propagation varies with the volatile content of the coal and the value of the air/coal ratio. The rate of flame propagation increases as the ash content of the fuel is reduced and as the ratio between volatile matter and ash increases. A bituminous coal with a volatile content of 30% and only 5% of ash can have a flame speed of 13.7 m/s whilst a fuel with only 15% volatile matter and 5% ash will have a flame speed of approximately 4.27 m/s. The flame speed rises to a maximum as the air/coal ratio increases and falls off as the ratio increases further.

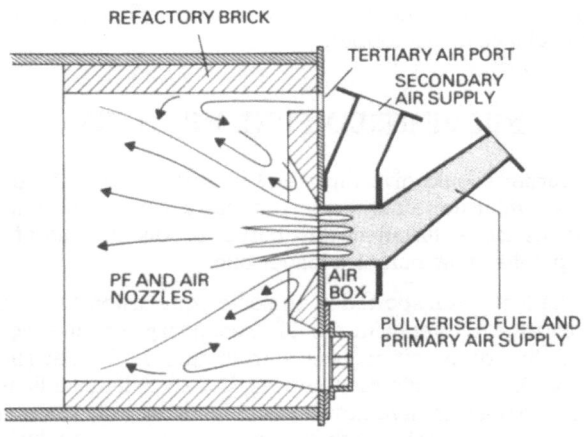

Fig. 22 Parallel flow pulverised fuel burner

The velocity of flame propagation of the air/fuel mixture in the burner is also important. The burner will backfire if the flame speed for the mixture is greater than the speed at which the mixture passes through the burner. If the flame speed is much less, the flame will lift off from the burner and ignition may be lost. The rate of flame propagation also affects the ease with which a PF burner can be lit. Where the correct mixture for normal running provides a fast rate of flame propagation, provision is made in the burner for some of the carrying air from the air/fuel stream to be extracted for a short period during 'light-up'.

Pre-heated primary and secondary air are useful when burning PF but primary air should not be pre-heated when pulverised pitch is used. When a bituminous coal is burned, care must be taken that the primary air temperature is not so high that the volatiles begin to distil off before the coal passes through the burner tip. Temperatures of up to 370°C are possible, however, when firing semi-anthracites and anthracites.

ASH HANDLING

As the fuels used for pulverising generally have high ash contents, mechanical methods of ash removal are usually needed and PF furnaces vary in type according to the way this is done. When ash forms and solidifies inside the furnace, it is known as a dry bottom furnace. In a slag type furnace, the ash is deliberately kept above fusion temperature and tapped from the furnace as a slag.

As PF burns in suspension in the air streams, not all the ash is deposited inside the furnace. Some escapes, through other passages in the boiler and eventually reaches the chimney top. For this reason, it is illegal to operate a PF furnace without a grit arrestor.

Pulverised fuel used in shell-type boilers burns in the furnace tube and ash is deposited further along the tube. Steam blowers are generally used to blow the ash forward into the secondary flues when it can be measured. There are very few if any shell boilers now fired on PF.

Chapter 9

LIQUID FUELS

The word 'oil' is generally used in this chapter to avoid repetition but all types of liquid fuel are covered.

DELIVERY, HANDLING AND STORAGE

Liquid fuel is delivered in tankers and great care must be taken to avoid spillage when discharging it into storage tanks. The tanks must be provided with vent pipes open to atmosphere and fitted with draw-off cocks for both sludge and water. All liquid fuels contain a small amount of water. When a petroleum fuel is heated, the water collects at the bottom of the tank. It collects at the top of the tank if a coal tar fuel is heated. The water must be drained off at intervals to prevent interference with the supply system and the burner. A dip stick must be used from time to time to verify the quantity of oil in the storage tanks and to check the reliability of permanently installed content gauges. The appearance of the dip stick could give a guide as to excessive sludge built-up since the lower end of the dip stick would show the characteristic sticky conglomeration of the sludge.

The heavier fuels have to be heated before they can be pumped easily from the storage tank to the boiler house. The storage temperatures for the various grades of liquid fuel are shown in Table 1 (p. 7). Storage tanks are usually heated by means of steam coils or electricity and often only the required flow is heated by outflow heater and not the complete contents of the tank. If steam is used for heating, the steam traps must be in good order. It is very important to ensure that the oil level in the tank is never below either the thermostat or the heater.

Bulk oil tanks are usually placed within a bund wall to prevent spilled oil escaping and becoming a fire risk. Safety precautions are dealt with in Chapter 13.

The pipes which carry the heated oil from storage to the burners should be lagged and require electrical or steam trace heating inside the lagging to prevent the oil losing heat between storage and the secondary heater. With 35 s oil such precautions are usually unnecessary.

Filtering of oil is of the utmost importance to remove foreign matter and this will accumulate and interrupt the oil flow unless the filters are regularly changed over and cleaned with gas oil or paraffin. Petroleum-based and coal tar-based fuels must not be mixed together or a thick residue will form and block the whole system.

FUEL OIL SYSTEMS

1. GRAVITY FEED
   This system can be used when only one or two burners are involved. The oil is delivered into an overhead tank from a tanker or by a transfer pump from storage. A pipe conveys the oil from the tank to the burner and the oil must be thin enough to flow freely.

2. CIRCULATING RING MAIN AT ATOMISING TEMPERATURE
   Oil from the storage tanks passes via the outflow heater to the pumps and final heaters, where the temperature is increased to ensure the correct atomising temperature at the burner (see Table 1). The heated oil is then circulated through a main from which tappings are taken to each of the burners. Surplus oil is returned to the inlet of the final pre-heaters. The pressure in the ring main depends on the type of burner and is controlled by a spring-loaded relief valve in the return loop. See Fig. 23.

3. CIRCULATING RING MAIN AT STORAGE TEMPERATURE
   Oil is pumped to the ring main and supplied to the burners through individual final heaters. Except when gas oil or CTF 50 is used, the main should be fitted with suitable electrical or steam tracing and lagged. Whilst it is not customary to provide a source of heat in storage tanks or around mains

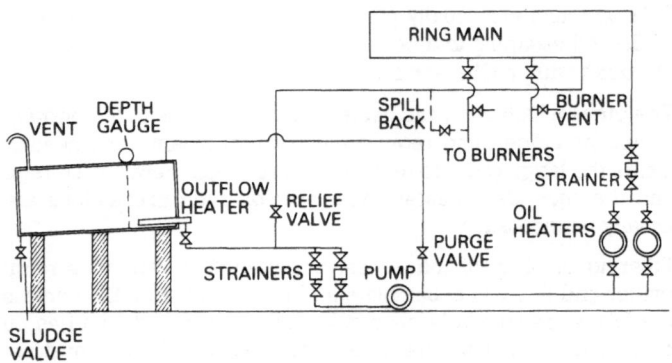

Fig. 23 Ring main system

containing gas oil it has been known for flow to be impeded under extreme conditions of cold or exposure and in these circumstances a facility for the application of moderate heat might be considered desirable.

OIL HEATERS AND FILTERS

Most of the liquid fuels used in boiler plant require conditioning by pre-heating and the filtering out of solid matter before being atomised into fine droplets and burned in the furnace.

All liquid fuels, other than gas oil or CTF 50, need heat to reduce their viscosity to a point where atomisation takes place at the burner and this is provided by secondary heaters, usually placed between the pumps and the burners.

Additional filters are usually placed in the system between the final oil heaters and the burners because foreign matter is more easily dropped out when the oil is warm. Filters need regular cleaning to prevent interruption to the fuel supply to the burners.

LIQUID FUEL BURNERS

The purpose of a liquid fuel burner is to atomise the fuel to the point where ignition takes place and combustion continues until the fuel is completely burned out. Some of the various types of burners used in conjunction with boilers are described below.

1. PRESSURE JET

The oil is supplied under pressure to a nozzle with tangential slots, designed to give spin to the oil before final discharge. A film is formed which breaks down into a fine spray when the oil emerges from the jet. All combustion of air is supplied around the burner (see Fig. 24).

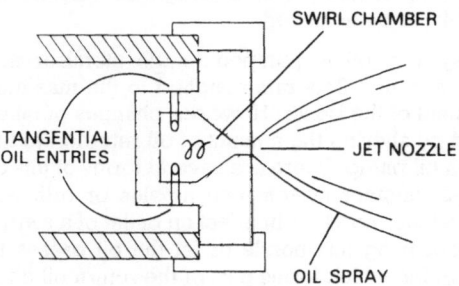

Fig. 24 Pressure jet burner

Oil pressures usually range from 6.8 to 17 bar but in some cases higher pressure are utilised. Such burners can have interchangeable nozzles having different orifice sizes to accommodate a range of boiler load characteristics and these are changed manually at the discretion of the operator or the Engineer in Charge.

This is necessary because of the limited 'turn-down' possible with pressure-jets. The range of load with a single tip being of the order of 2:1.

Fully automatic pressure jet burners were developed for the Shell boilers and particularly for the so-called Package boiler.

The compact pressure jet arrangement consists of a motor driven fan, providing combustion air and also geared to or connected to an oil pump. Within the same general external sheet metal casing are housed transformer ignition electrodes and sequence controller for the ignition and shut down of the burner. All units are prewired, easy to install and being completely automatic require very little attention. Such burners are available as On/Off, High/Low/Off or Fully Modulating Units and these terms apply to the control system rather than to the principle of the burner design as such. These units will be discussed in Chapter 12.

91

The straightforward pressure jet burners are associated with an Air Register and forced or induced draught, or a balanced draught system can be incorporated.

2. SPILL JET BURNERS
Conventional pressure jet systems have a limitation of a low turn-down ratio, and the spill jet system enables turn-down ratios of 4:1 to be achieved.

In this system, oil is pumped to the atomiser at constant pressure and at a flow rate matched to the maximum output requirement of the boiler. However, changes of oil output are achieved by altering the amount of oil returned to the suction side of an oil pump. There are several forms of jets or nozzles and in yet another system two nozzles or orifices may be used. The inner nozzle is in effect an outlet of a swirl chamber and after leaving this nozzle or jet the oil passes through a second orifice which forms part of the return oil circuit.

Even with a wide turn-down range, the degree of atomisation is quite constant although perhaps somewhat coarser than with the standard pressure jet burners. Variations in flame width may occur, resulting in difficulties when applied to certain types of boilers.

3. STEAM OR AIR ASSISTED PRESSURE JET ATOMISERS
Oil is supplied to a pressure jet nozzle and is delivered as a spray to a mixing nozzle where air or steam is admitted through a series of orifices as an additional means of securing fine atomisation.

The steam or air quantity used is less than 1.0% of the steam generated or less than 1.0% of the combustion air. A turn-down ratio of 6:1 is possible and the system enhances the admixture of combustion air due to its high energy. In general combustion is cleaner and less stack solids are produced.

4. HIGH PRESSURE AIR OR STEAM BURNER (BLAST ATOMISERS)
Oil is supplied at a controlled rate to a nozzle where air or steam at pressures over 1 bar meet the fuel at an angle and causes atomisation. Only a small proportion of the total air is provided via the burner as primary atomising air, the rest being supplied as secondary air around the burner. In the steam burner all the air is supplied around the burner, similar to the method of secondary air admission.

5. THE MEDIUM PRESSURE AIR BURNER

This uses the same principle as the High Pressure System but the air pressure applied ranges from 70 mbar to 1 bar.

6. LOW PRESSURE AIR BURNER

This uses the same principle as the foregoing but the air pressure is supplied from a fan at between 27 and 103 mbar. From 25 to 100% of the air passes through the burner. (See Fig. 25.)

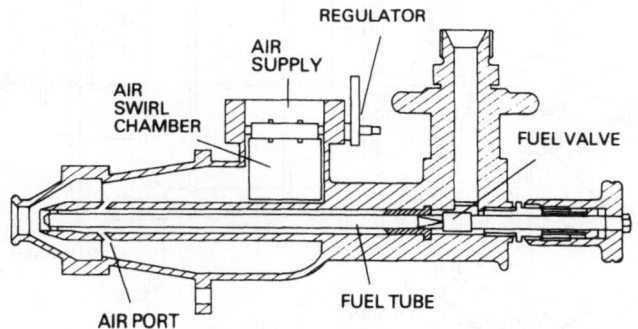

Fig. 25 Low pressure air burner

In most applications the burners are associated with Air Registers.

7. ROTARY CUP BURNERS

Oil is fed onto the inner surface of a rapidly rotating cone, or cup, where it is finally thrown from its lip and its breakdown assisted by an air blast fed around the periphery of the cup. Both the rotation of the cup and the fan is achieved by one motor and all are housed together with the essential ignition and fuel air ratio controls in one external casing. This gives a Packaged assembly in the same way as for fully automatic pressure jet burners. Secondary air is applied either by means of an Induced or Forced Draught fan. A good flame profile is achieved and the burners are relatively insensitive to changes in viscosity of oil when a metering pump is used. The burners tend to be noisy. Rotary cup designs are available as fully automatic, semi-automatic or hand controlled units and various principles of automatic control are available. (See Fig. 26.)

93

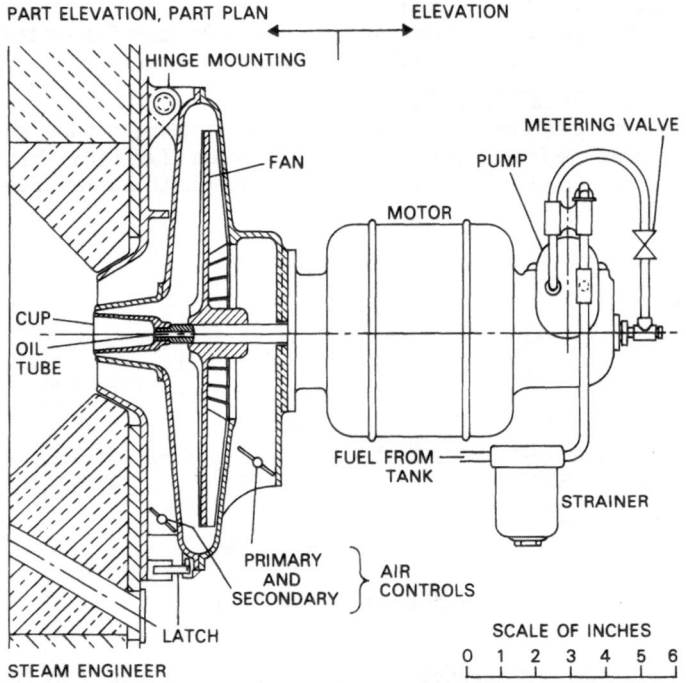

PART ELEVATION, PART PLAN | ELEVATION

HINGE MOUNTING

FAN

MOTOR

PUMP

METERING VALVE

CUP

OIL
TUBE

FUEL FROM
TANK

STRAINER

PRIMARY
AND
SECONDARY } AIR
CONTROLS

LATCH

SCALE OF INCHES
0 1 2 3 4 5 6

STEAM ENGINEER

Fig. 26 Sectional view of motor-driven rotary cup burner

8. EMULSIFYING BURNERS

The oil and some of the combustion air are pre-mixed at a
stage before the actual burner.

Sectional boilers are usually fired by self-contained automatic
pressure jet burners but for the smaller sizes the following
types are used.

9. VAPORISING BURNER

Vaporising burners which use a domestic grade of paraffin
are of four general types:

(i) The perforated sleeve type, where the burner consists of
one or more narrow vaporising troughs into which the oil
is fed through a constant level device. The sides of the
troughs are surmounted by perforated metal walls,

94

through which combustion air enters to mix with the oil vapour. Combustion occurs partly above and partly between the sleeves.

(ii) The pot type, where the oil is vaporised in an open topped pot and combustion air enters through perforations in the sides. The oil is supplied through a constant level device, generally by gravity, and the flame normally stabilises above the top of the pot.

(iii) The rotary vaporising type, either vertical or horizontal in which the oil is thrown by a rotating cup onto a vaporising surface over which combustion air is supplied by a fan.

(iv) Gasifying burners, in which oil is fed under pressure, or by gravity, into a heated vessel where it is gasified. Except at light-up, the vessel is heated by the oil itself. Combustion air can be preheated before passing through the burner.

There are **Advantages** and **Disadvantages** with all these burners. The pressure jets require fairly low power consumption for pumping, respond to automatic control and although the turn-down is usually low it can be as high as four to one in some types. The jets, especially of the manual gun-type associated with air registers, require regular and frequent cleaning.

In all air or steam pressure burners, the turn-down ratio can be as high as four to one or better, even ten to one in very large medium and high pressure burners. The flame shape readily adapts to the furnace and atomising air is easy and inexpensive to provide. Power costs are higher than with pressure jet burners, however, and the air trunking with the large sizes of low air pressure burners is cumbersome. Steam jet burners have a disadvantage in that they cannot be used on a boiler until steam is available. The spinning cup or rotary burners have no jets to clog up and give a reasonable turn-down ratio. Vaporising burners can be noisy and, in some cases, need maintenance for the fan motor. They often operate on gravity flow and do not need an oil pump. They are simple in operation and low in cost.

In the low, medium and high pressure air atomising and emulsifying burners, part of the combustion air is admitted to atomise the fuel. Secondary air needed to complete combustion is introduced into the furnace at the proper place and at the appropriate rate for the amount of oil being used through an air director, or register, surrounding the burner. Pre-heated air is helpful but primary air for atomising the fuel should not be hotter than the fuel itself. Secondary air can be at a higher temperature. The air director consists of a shutter or louvre, with a number of directional vanes surrounding the burner. The secondary air passing through it can be controlled in quantity and direction and is given a swirling movement to help produce rapid and complete combustion.

The burner and director are mounted into a refractory brick ring known as the quarl. When the burner is in use, the quarl becomes hot and radiates heat back to the cone of atomised oil and air, thus helping to stabilise the flame and maintain ignition. When a burner is cleaned, great care must be taken not to damage the quarl or an unstable flame may result which will impinge on the furnace walls and cause smoke and soot deposits. A damaged quarl or one that has eroded edges, should be repaired or replaced without delay.

BURNER MAINTENANCE

Burners can easily be damaged during cleaning but must be cleaned at frequent intervals to remove small amounts of carbon or gum which form on tips and may prevent correct atomisation. An oil burner withdrawn for cleaning must be stripped down according to the manufacturer's instructions and soaked in diesel oils or paraffin. If the burner uses one of the coal tar fuels, CTF50 or creosote should be used for cleaning. When the solids have softened the burner parts should be cleaned with a soft rag free from anything that can scratch the metal. The remaining deposits can then be removed with a very soft metal or wood scraper, anything hard will undoubtedly damage the surfaces. Some of the burner manufacturers provide special 'dollies' of soft metal for cleaning out the small jets. These are of the correct size and shape for the plant concerned and no effort should be made to force through a wrongly sized dolly or serious damage will result. If any part of a burner becomes damaged or worn, or an orifice becomes deformed, a replacement should be obtained.

CARE OF BURNERS
WHEN SHUTTING DOWN

If possible the burner should be removed whenever a boiler is shutdown even if only for a short period, to prevent radiation from the furnace overheating the burner. This can cause small amounts of oil to carbonise and may lead to deposits of carbon and solid matter. The opportunity should also be taken to clean the burner whenever it is removed. If it is not possible to remove the burner from the boiler it is possible in some cases for a radiation shield to be inserted between the burner and the furnace.

Some of the compact self-contained pressure jet or rotary burners can be swung away from the boiler front for essential maintenance.

CARBON DEPOSIT

Carbon is sometimes found inside the furnace on the sides of the furnace tubes and the refractory quarl for the following reasons:

(a) The air supply is wrong or not properly directed.
(b) The burner needs adjustment or modification.
(c) Atomisation is imperfect because of damage to the nozzle or quarl.

The location of the deposit may be a guide to the cause. If it forms on only one side of the furnace tube, the burner is probably off-centre and needs re-alignment, or the burner jet orifice may be damaged.

If a rotary cup is not centrally situated in relation to the primary air annulus surrounding the lip the flame can be deflected towards one side or the other of the furnace tube.

Soot or smoke can form if atomisation is poor, combustion slow, or the flame is chilled by cold air or by impinging on a cold surface.

Chapter 10

GASEOUS FUELS

Town Gas obtained from the carbonisation of coal is no longer used to fire boiler plant, except perhaps in association with plant operated by Coke Producers, e.g. the British Steel Corporation, National Coal Board etc. The usual calorific value of such a gas is about 18.6 MJ/m^3 at 15°C and atmospheric pressure.

NATURAL GAS

Natural gas, the product of gas wells or a by-product of oil wells, is mostly Methane and has a calorific value of around 38.2 MJ/m^3.

LIQUID PETROLEUM GASES (LPG)

Liquid petroleum gas produced during the refining of crude petroleum oils, is heavier than air and on this account requires even more careful handling than town or natural gas. It is supplied in liquid form under pressure in cylinders or tanks. The actual gases in this class are commercial butane and commercial propane which, in addition to being used in boilers or furnaces, frequently provide the gas flame for igniting liquid fuels.

COMBUSTION CHARACTERISTICS

Gases need to be mixed with combustion air before they can burn and the air/gas mixture has a characteristic flame speed between the upper and lower limit of flammability. This is the speed at which the flame moves along a tube containing the mixture when the waste gases of combustion escape to atmosphere. Flame speeds depend on the air/gas ratio and vary from one gas to another. There are also upper and lower limits to the air/gas ratio outside which a flame cannot be propagated and in many burners air/fuel ratios are automatically adjusted throughout the range of the burner output.

These limits are shown in the table below.

Type of Gas	Flame Propagation Air/Gas Ratio Limits
Town Gas	5 – 30
Natural Gas	5 – 15
Commercial Butane	1.9 – 8.5
Commercial Propane	2.4 – 9.5

The flame speed of the air/gas mixture also determines the stability of the flame at the burner tip. If too high, the gas may light back into the burner; if too low, the flame will lift off the burner tip and ignition may be lost. The tendency for the flame to lift off is much greater for natural gas than town gas. It is therefore most necessary to modify or change the burner completely when converting from one gaseous fuel to another.

TYPES OF BURNERS FOR USE WITH NATURAL GAS

Low Pressure burners operate over the range 2.5 – 10 mbar gas supply pressure, whilst high pressure burners operate over ranges 12 – 175 mbar.

The low pressure burners, usually of the multi-jet type are used for low heat input, the positioning of the jets being compatible with the type of appliance, that is, the boiler to which they are fitted. Combustion air is either naturally or fan induced and the systems are sometimes known as 'free flame' burner units. Figure 27 shows a typical natural draught injector type.

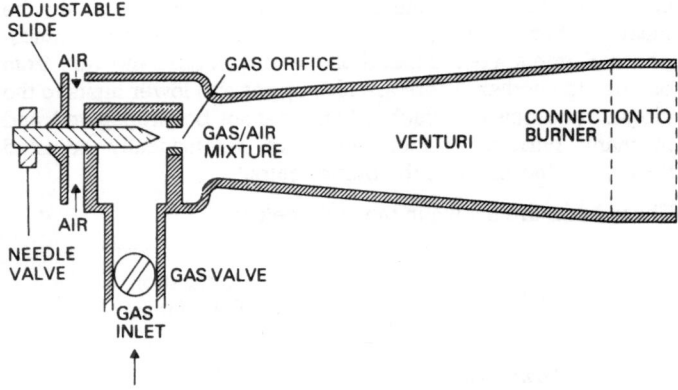

Fig. 27 Atmospheric injector burner

HIGH PRESSURE BURNERS

These are used for large heat input and are further classified according to the arrangement for mixing gas and combustion air. The following are some examples of high pressure burners.

GAS RING TYPE.

The air is controlled by an adjustable register, and this alters quantity, velocity and direction of flow and exercises a similar function to air registers used with pressure jet type oil burners. The gas supply, from a ring, is positioned between the register and the furnace wall surrounding the burner opening. Gas is directed from the ring across the air stream towards the burner opening, the mixture of gas and air entering the furnace through a throat or guard. The system can readily be adapted for dual fuel firing, although it necessitates the withdrawal of the oil burner when firing with gas. It can usually be seen on the older Economic, Lancashire and Water Tube Boilers.

Centre-diffusion tube burners are also used. The gas is supplied to the centre of the burner and the gas burner unit is sometimes made large enough to contain concentrically a mechanical atomising liquid fuel jet which is also again retracted when firing gas only. Air registers and a diffuser are provided to ensure the rapid mixing of gas and air, thus giving good and rapid ignition with a minimum of excess air.

100

Turbine gas burners driven by the pressure energy of the gas supply are also available and are similar to the rotary or spinning cup oil burners in terms of maintenance of fuel/air ratio at various fuel input. Again such burners can be made available for dual fuel burners.

DUAL GAS/OIL BURNERS.

As their name implies, they burn either oil or gas or both. It is impossible to describe here the various burners of this type and, when encountered, the manufacturers' instructions should be followed with great care. In most of them, however, the gas is fed into the refractory quarl surrounding the oil burner or spinning cups. Some of the combustion air can be supplied through the quarl and the remainder, which is controlled both in quantity and direction by louvres, enters through the burner mounting. When using oil, the gas supply is shut off and the burner operated in the usual way. With gas, the oil burner or its nozzle must be protected from overheating by the gas flame or radiation from the furnace. The method of protection, however, must not interfere with the direction of the air/gas streams and destroy the flame shape. (See Fig. 28.)

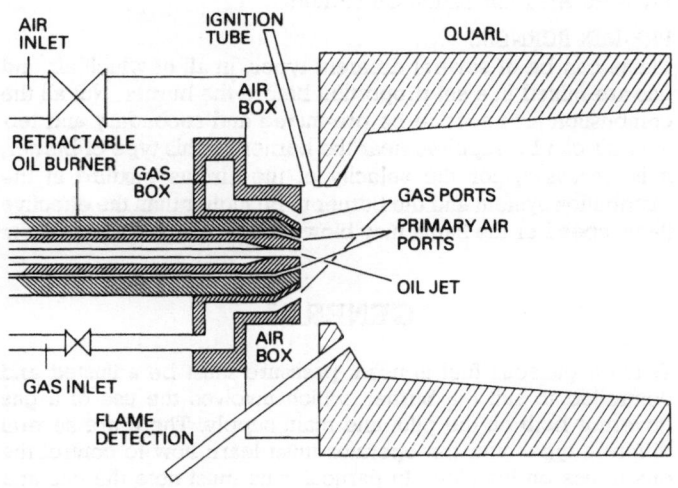

Fig. 28 Dual gas/oil burner

When both fuels are used simultaneously, the air supply is much more difficult to control and the burner will need frequent cleaning to remove the carbon caused by overheating.

On the larger water-tube boilers Tangential burners are used when dual firing of gas and pulverised fuel is required. Gas or pulverised fuel is directed from the corners of the boiler across a small circle at the furnace centre, whilst a forced air supply is directed across the gas nozzles.

Some shell boilers are fired by **low pressure blast burners**, the gas being supplied at pressure between 5 mbar and 20 mbar whilst cold or pre-heated air is provided at pressures up to 25 mbar.

The gas and air are fed into a refractory block or quarl similar to that used for oil burners and both are controlled by separate valves which may be coupled together to ensure a constant fuel/air ratio. In such cases, the pressures and temperatures of the air and gas supplied must be very strictly controlled. The low pressure blast burner is also associated with the use of air registers when only part of the combustion air passes through the burners whilst the remainder is supplied as secondary air through the air register. Again similar in application to Low Pressure Air Atomisation Oil Burners.

PRE-MIX BURNERS
These can be of several different types, in all of which air and gas are mixed in a set proportion before the burner. Not all the combustion air needs to be pre-mixed and secondary and tertiary air can be supplied near the burner. In this type of burner, it is necessary for the velocity of the air/gas mixture in the distribution system and the burner to be higher than the effective flame speed or the flame may blow-back or cause an explosion.

GENERAL

When a gaseous fuel is used, pressure must be adjusted and controlled to suit the burner, which involves the use of a gas governor both on the pilot and main supply. There are several different types and the operator must learn how to control the one in use on his plant. In particular he must note the gas and air pressures which are necessary at the burner and relate these to information supplied by accurate pressure gauges, which in the case of gas supply must be installed downstream from the governor. Sufficient pressure tappings should be made available at the side of automatic valves and governors so that in the event of burner ignition problems the correct operation of such valves and governors can be verified. The governor should be installed in a position where it can be kept clean and will not become overheated. Some have diaphragms which can deteriorate in use.

Gas burners need correct maintenance and they are as vulnerable as oil burners to damage by careless handling or the use of the wrong cleaning tools. The manufacturers's instructions should, therefore, be carefully followed. If damage occurs, a change of combustion characteristics and an alteration to the flame shape will probably result.

In a similar manner to oil burners, gas burners are built into automatic units where the primary source of ignition is an electric spark or a pilot gas flame. If ignition is lost, the automatic control shuts down the main gas supply and purges the system of gas before attempting to re-establish the flame. When the flame cannot be relit, the control gear locks out to safety. If this should happen, the operator can re-set the automatic control or change to manual operation but he must never alter the timing sequence of the automatic controls. Before the pilot flame is relit or the electric spark checked, the dampers must be opened to the fullest extent and the system thoroughly purged of gas. This latter check operation is as equally applicable to oil firing methods and to start up procedures for coal fired systems. Safety Precautions are dealt with in Chapter 13.

Figure 29 shows an arrangement of burner, governor and control valve system.

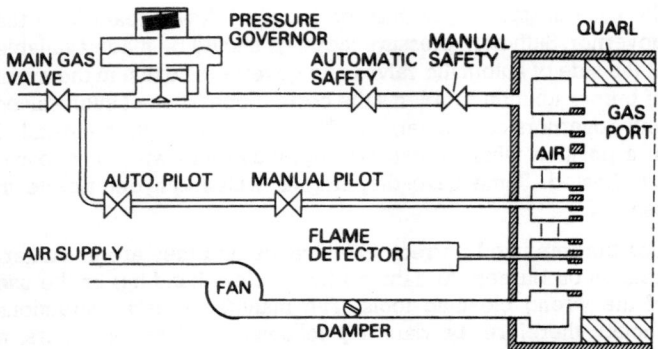

Fig. 29 Automatic gas burner

104

Chapter 11

CENTRAL HEATING: WHAT THE CARETAKER SHOULD KNOW

Small central heating boilers are usually in the charge of a caretaker or porter who acts as part-time boiler operator and is responsible for maintaining heating and hot water services. This chapter is specially included to provide guidance on the operation and maintenance of this type of plant. Some sectional boilers are still hand fired with coke or anthracite and this is the type described, though much of the information is relevant to newer systems, especially that concerned with care and maintenance.

PLANT OPERATION

Operation of a heating system depends on weather conditions. With sectional boilers, it is necessary to clean the fire, recharge it some time before the building is occupied, again at mid-day, and finally in the evening. A slower fire will usually maintain the building temperature during the afternoon. The boiler can be put out of commission during weekends if the building is unoccupied, provided the weather is mild, but it must be kept in operation if there is likelihood of damage to the system by freezing.

Automatic firing of solid fuel, gas, or oil, is controlled by thermostat or time switch and the operator should follow the manufacturer's instructions regarding this equipment.

DRAUGHT CONTROL

Primary air passing through the fire is adjusted by the damper on the door of the ashpit. Secondary air is controlled by another damper, generally in the form of a slide or rose, on the firing door. Some boilers have only fixed openings on the firing door. The total pull on the boiler is controlled by the flue damper, which can be of the sliding plate or butterfly valve type. The damper should not close the flue completely and about 10% of it is usually cut away so that the boiler can be banked with the damper in the closed position.

A check-draught damper or door is sometimes fitted on the chimney side of the flue damper to reduce chimney draught by allowing air to enter and cool the flue gases. A check-draught or plate damper in the chimney itself is essential if it is a high chimney or there is too much uncontrolled draught. A draught stabiliser will ensure a steady draught whatever the outside temperature or strength of wind. Draught should not fluctuate and must always be the minimum required to meet the load. Check dampers which admit cold air to the chimney are not advisable when high sulphur fuel oil is in use.

If the boiler is fired every five hours, the depth of the fuelbed will fall between firings from about 46 to 15 cm and the amount of secondary air needed will change a great deal during this time. For the best results, the secondary air damper opening should be adjusted for the average depth of the fuelbed. The damper setting should give the maximum CO_2 at half-time between firings and, if a CO_2 recorder is not fitted, a good guide is to allow from 560 to 1500 mm^2 of secondary air opening for each 23 kg of solid fuel burned per hour.

Air which leaks into the boiler will cause waste. Doors cannot be made completely air tight, but the gap between the door and frame should be restricted to the thickness of a postcard. Doors which are distorted or badly fitting should be overhauled or renewed. Air leakage also adds to the risk of corrosion by cooling the system below the acid dewpoint of the flue gases.

REPAIRS

A common cause of failure to control the heat output is air

leakage into the ashpit between the boiler and its foundations. Few central heating boilers have closed ashpits and expansion and contraction of the side walls causes the cement to crack and air to be admitted. The ashpit should be regularly examined and defects made good.

The firing and clinker doors usually have removable linings which must be replaced when burnt or distorted.

All boiler repairs should be made by a skilled craftsman.

AIR INLET TO BOILER ROOM

If the boiler house has insufficient ventilation, combustion will be affected and the atmosphere may become poisonous. If it is usual to keep the door closed, fresh air must be provided by proper and adequate ventilation, preferably at high and low level.

A good general rule is to allow $1m^2$ of ventilation opening for every 1000kW of boiler capacity (oil or coal). Approximately 70% of this opening should be provided at low level. For gas-firing the area should be increased by 50%.

TOOLS

Suitable tools should be provided and maintained in good condition. The following are suggested:

(a) A shovel of the right size for the fire door opening. One with deep sides will be needed if coke is burned.
(b) A slice bar, which has a flat blade with a light chamfer at the end so that it can be pushed along the firebars to detach the clinker.
(c) A hoe, rake or scraper for removing ash and clinker, when cleaning the fire.
(d) A flue brush of suitable size and shape. The head should be renewed when necessary and a chimney sweep-type brush with steel bristles will be useful.
(e) A pricker bar for clearing the spaces between the fire-bars from underneath.
(f) A steel barrow, or other effective means of removing ash

and clinker from the boiler house immediately after fire cleaning.

FUEL

The fuel should be allowed to dry under cover before use. Coke, anthracite, or Welsh smokeless coal are usually burned in this type of plant, the fuel size depending mainly on the grate area. Except for the small boilers, there is considerable latitude in choice. If the fuel contains too much dust and pieces below 3mm in size, clinker will be troublesome. Fuel selection will depend on the advice of the boiler manufacturer but the following table is a general guide.

Grate area m²	Heat Output kW	Fuel size limit mm
0.1–0.5	15–60	25–50
0.5–1.0	88–220	25–50
		50–75
above 1.0	above 220	50–125

OPERATING PROCEDURE

START OF HEATING SEASON

The following points should be checked before starting up:

(1) That the system is filled with water. If an altitude gauge is fitted, the black and red pointers should coincide.
(2) That the ball valve in each feed and expansion tank is working properly.
(3) That the stop valves (those close to the boiler on flow and return) are open.
(4) That the relief valve is free on its seat.
(5) That the system is free of air, by checking air cocks on radiators and pipes.
(6) That valves on any circulating pumps are open and the pumps in working order.

LIGHTING UP

(1) The ashpit should be clean and the firebars free of obstruction.

(2) Ashpit and flue dampers are then opened and the check-draught damper closed.

(3) The circulating pump is started.

(4) The fire is lit with the usual kindling material and fuel is added gradually when it is well alight.

(5) The air vents should be checked when the system is warm.

FIRING

(1) Before disturbing a banked fire, and prior to opening the firing door, the dampers should be opened for some minutes to clear the flues of unburned gases.

(2) A level bright red fire should be built up over the grate.

(3) Special attention should be paid to fuelbed thickness. If too deep, the boiler heating surface will be shielded from direct radiation and unburned gases will be lost. The fuelbed should not, however, burn so low that excess air passes through and cools the heating surfaces.

(4) The fuelbed depth should be kept within the following limits:

Size of fuel	Depth of bed	
mm	Coke mm	Anthracite mm
25–50	200–380	150–250
Over 50	300–450	200–380

(5) The firebox should not be more than three-quarters full if coke is used or more than half full with anthracite and these limits must be observed when banking overnight.

(6) After firing, sufficient secondary air must be introduced over the fuelbed to burn the gases.

(7) The boiler must not be forced by excessive draught. Intense heat damages firebars and causes clinkers.

109

(6) The top of the fire should not be disturbed or clinker formation will be increased.

CONTROL

(1) The burning rate should be controlled by chimney and ashpit dampers; the most suitable positions have to be learned by experience.
(2) The flow temperature must be selected according to outside shade temperature and the following table is a guide.

Morning outside temperature		Boiler flow water temperature	
°C	°F	°C	°F
10	50	49	120
7	45	57	135
4.5	40	66	150
2	35	74	165
−1	30	82	180

If a higher flow temperature than indicated becomes necessary, the reason should be investigated.

When operating at low flow temperatures, the possibility of an increase in corrosion must not be ignored, especially when using high sulphur content fuel. If there is more than one boiler, it may be prudent, during light load periods, to reduce the number of boilers in use. Radiators under the direct control of the caretaker should be shut off and a high water flow temperature should be maintained through a reduced heating surface.

(3) The water temperature should not be higher than weather conditions warrant. If a building is overheated by only 1°C about 9% of fuel is wasted.
(4) Where heating is by unit heaters or hot air cabinets, the minimum temperature given in the instructions, usually between 60 and 71°C should be maintained.
(5) The flow temperature of a domestic hot water service boiler should not usually exceed 60°C (140°F) nor fall below 43°C (110°F).

(6) It may be necessary to visit the boiler room to reset the dampers about 30 minutes to one hour after recharging the boiler.

(7) The fire door must not be left open.

FIRE CLEANING

(1) The fire should be cleaned when it has burned down to minimum thickness.

(2) The dampers must be opened while cleaning to clear away the fumes.

(3) A thorough, but quick, cleaning is essential and one side of the fire should be cleaned at a time. The good fuel is moved to one side, to expose the ash and clinker to be removed, and then moved back to the clean half of the fire for the operation to be repeated on the other side of the grate.

(4) The clinker door should be used for cleaning. This prevents the door being broken, as can happen if the firing door is used for this purpose, and avoids an increase in clinker formation by clinker being pulled up through the fire.

(5) To prevent an unpleasant and corrosive atmosphere, ash and clinker should not be quenched in the boiler house but removed as quickly as possible into the open air.

GENERAL CLEANING

(1) The boiler flues must be cleaned and the water surfaces in the firebox scraped at least once a week.

(2) The boiler room must be kept clean, tidy, and free from dust, and there should be easy access to the damper at the back of the boiler.

BANKING

(1) In some weather conditions, fuel can be saved by banking early in the afternoon or evening.

(2) Depending on the type of fuel, the fire should be built up slowly until the firebox is three-quarters full of coke or half full of anthracite or Welsh steam coal.

(3) The dampers should be set so that the fire remains alight

during the night and the temperature is sufficiently high in the morning to avoid forcing the boiler.

(4) The check-draught damper should be opened and then adjusted to reduce the chimney draught.

(6) Circulating pumps should not be left running.

MAINTENANCE DURING THE HEATING SEASON

(1) Air leaks at the base of the boiler, between boiler sections, at the joints of the flue pipe connection to the chimney, at badly fitting doors, or flue covers, should be sealed.

(2) The boiler must not be used to burn rubbish.

(3) Damaged lagging should be reported for repair.

(4) A standby circulating pump should be used alternatively week by week with the other pump.

(5) Defects or irregularities should be reported.

(6) Wornout firebars and firing tools should be replaced.

(7) Dampness in the ashpit, or any sign of water leakage, should be reported without delay as it is often the first sign of serious trouble.

MAINTENANCE AT THE END OF THE HEATING SEASON

(1) If the boiler is out of service for a prolonged period, unburned fuel, ash, clinker and loose firebars should be removed and the flues and chimney connection cleaned.

(2) Interior surfaces of the firebox and flues should be painted with lime wash to prevent corrosion.

(3) Unless the boiler is connected to a common flue used by other working boilers, all dampers, doors and flue-cleaning covers should be left wide open to allow air to circulate, and the main chimney should be swept.

(4) The system should not be drained, unless there is risk of freezing, as fresh water is likely to cause scale deposits.

(5) The flow and outside shade thermometers should be checked against a known accurate thermometer.

(6) The protective lining of fire and clinker doors should be replaced if burnt or distorted.

RECORDS

(1) The quantity and type of fuel received and consumed should be entered in a log.

(2) The boiler flow and outside shade temperatures should be logged each day.

BREAKDOWN

(1) Water leaks in the boiler or heating system should be reported to the management without delay.

(2) If the leak occurs in the boiler, flow and return valves must be shut, the circulating pump stopped, and the boiler fire drawn.

(3) If the leak is in the heating system, the part affected should, if possible, be isolated and, if necessary, drained.

(4) If these measures are unsuccessful, the boiler fire must be drawn and stopcocks and valves to the feedwater tank shut or the ball valve tied up.

(5) If the circulating pump fails and the water temperature is likely to exceed 88°C, the pump must be by-passed and the fire drawn.

FROST PRECAUTIONS (manual systems)

(1) Boiler fires must be maintained at night and weekends and circulating pumps kept running.

(2) Radiator control valves should be kept open and inlets for fresh air behind the radiators closed, where fitted.

(3) If the fire dies out, it must not be relit without checking that no part of the system, including the relief valve, and open vent pipe, is frozen.

Building services have become more sophisticated, the aim being to attain improved comfort conditions relative to a minimum use of energy. Therefore the older type sectional boilers are being replaced where site conditions permit, by more thermally efficient 'packaged' boilers. Such boilers incorporate automatic firing of either solid, liquid or gaseous fuels controlled by thermostat, froststat, time switch, and even combination systems where the start up time can be altered in relation to the prevailing outside temperature. System controls can also incorporate three way and four way mixing valves, which in addition to effecting room temperature control, reduce the risk of possible boiler corrosion when using solid and liquid fuel containing sulphur. It achieves this by ensuring that the boiler water temperatures remain at a sufficiently high level to minimise the condensation of acidulated moisture on the gas side of the flues. It is not uncommon to see boiler houses containing fully automatic oil and gas fired boilers, almost 'hermetically' sealed and using solid doors for security reasons. In these instances soot and dust build-up on flue passages has been as great as that experienced with solid fuel firing systems and it is little wonder that complaints are received regarding the inadequacy of the heating systems. There is a danger with smaller boiler systems for an out of sight out of mind mentality to develop.

It is important to note also that bad ventilation can cause build-up of dangerous fumes which, especially with gas-firing, may not be readily detected by smell. Many serious accidents have resulted because of this.

Chapter 12

AUTOMATIC CONTROL AND INSTRUMENTS

AUTOMATIC CONTROL

Automatic control plays a big part in relieving the tedium of labour and, when installed in a boiler house, can enable a single operator to look after a large number of boiler units. Mechanical devices for fuel handling also mean greater cleanliness and better working conditions.

Maintenance, adjustment and installation of automatic control is a specialised field, the principles involved ranging from the very simple to the extremely complex. Whatever the system, the automatic device starts by measuring the variable to be controlled and comparing it with a set standard of temperatures, pressure, water level, steam or water flow, or draught and often combinations of these parameters.

The difference between the measured and set point initiates corrective action until the two points are brought together. At this stage, when there is no difference between them, the automatic device falls into neutral. To give a simple example, when water in a tank is controlled by ball valve, the floating ball measures the level and keeps the valve open until the water in the tank reaches the required point.

Automatic systems must be maintained and adjusted by a skilled technician. The operator should only be required to check, by reference to the instruction book, that the system is in accurate working order and report any defects to the responsible engineer. Logging of data is just as essential in the efficient operation of an automatically controlled boilers as it is for manually operated boilers. Uncharacteristic deviations from the norm could

115

give an indication that the control system is not functioning properly.

Total reliance on the automatic control system may breed an attitude of indifference towards the plant and signs of possible danger ignored. It is just as important to visually check water level in the gauge glass as hitherto. Whilst more appropriate under the general heading of Safety the following is a general observation made as a result of discussions with students attending NIFES Boiler Operator Courses.

In the days of hand firing and to some extent when using moving grates the steam safety valve lifted and blow off occurred. Often, this was too frequent and fuel savings were possible by keeping blow off to a minimum. On the credit side, so to speak, the operator at least knew his safety valve was working.

Especially in the case of oil, gas and underfeed stoker fired boilers the safety valves rarely, if ever lift of their own volition and the attendants at NIFES Courses had never seen the safety valve lift on their own account. When questioned further, 80% had only seen the safety valves blow off once/year during inspection and testing procedures. Such was the confidence in automatic control that it was assumed that the heat source would cut off when required and so prevent blow off and yet little regard was given to the vital and final safety back-up.

Automatic control in relation to boilers of all types serves three purposes:

(a) A safety function.
(b) A combustion efficiency function.
(c) Labour saving function.

SAFETY FUNCTION

When a burner is started, the first essential is for a flame to be established. A device is required which monitors the presence of a flame or heat source and this is known as a flamestat. The flamestat will cut off the supply of fuel if the flame fails to appear or goes out. In some oil fired systems no flame can be established until the oil is at or near its atomisation temperature.

Coincidental with the function noted above there should be an interlock which cuts out the whole system if the boiler outlet damper is shut and certainly if the water in the boiler is dangerously low.

Once the flame has been established the operation of the burner is taken over by a thermostat, pressurestat or flowstat or by combinations of temperature and flow or pressure and flow, with an overriding limit stat for temperature or pressure, dependent upon the function of the boiler. In a fully automatic system protection against any sort of failure must be incorporated into the control system. Ignition control follows a time sequence and BS 799 sets down the parameters required for effective control.

A hypothetical control for a large burner with pilot is shown in Fig. 30 but it must be borne in mind that interlocks are incorporated in the circuit for damper position, water level, electrical failure of a component, low oil temperature and limit stat. Normal operation of the burner is by way of the primary control setting e.g. temperature or pressure.

FUNCTIONAL CONTROL

By this is meant a control related to the attainment of a set pressure, temperature, flow and thermal efficiency of the boiler. Doubtless the boiler attendant has heard control referred to as On/off, High/low/off, or Modulating and it is necessary to define these terms, and in particular their effect on thermal efficiency. It is also possible that the attendant had heard the term proportional control and associates this in terms of fuel and air proportionability or fuel/air ratio. This is not the case; it is merely a control function term usually associated with fully modulating control where the heat input is controlled progressively in a band relative to the required set function.

ON/OFF CONTROL

The simplest example is where a steam pressure or hot water temperature control can be set to give a 'run or stop' signal to a single motor driving an underfeed stoker or oil burner. Usually this motor drives both the air and fuel feed mechanism, although with oil and gaseous fuel burners, a delay device is usually incorporated to keep the fuel valve shut to give a purging and

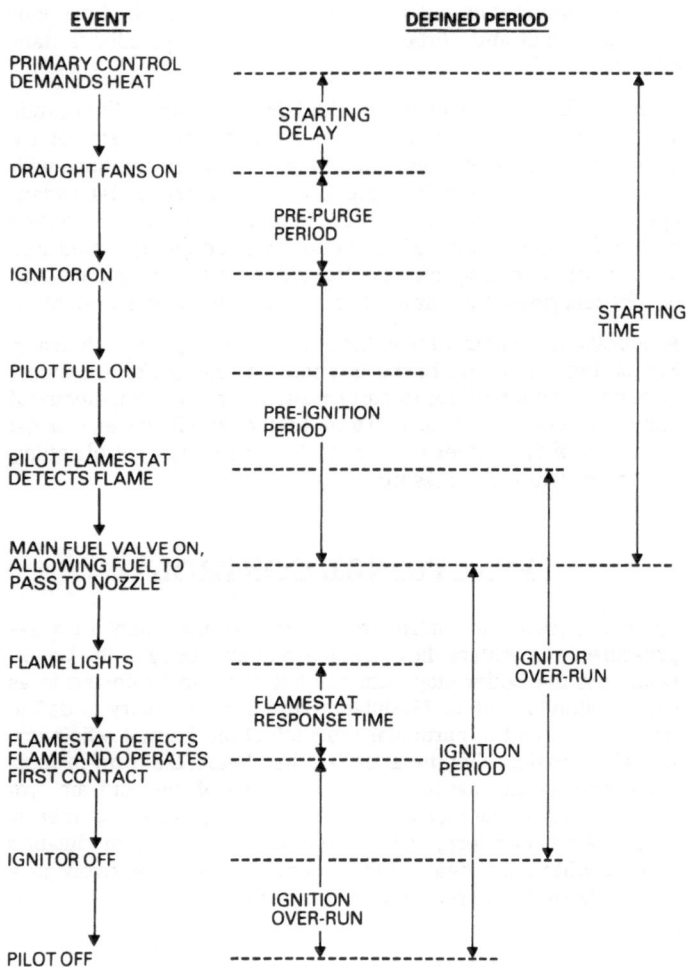

EVENT	DEFINED PERIOD

PRIMARY CONTROL
DEMANDS HEAT

STARTING
DELAY

DRAUGHT FANS ON

PRE-PURGE
PERIOD

IGNITOR ON

STARTING
TIME

PILOT FUEL ON

PRE-IGNITION
PERIOD

PILOT FLAMESTAT
DETECTS FLAME

MAIN FUEL VALVE ON,
ALLOWING FUEL TO
PASS TO NOZZLE

FLAME LIGHTS

IGNITOR
OVER-RUN

FLAMESTAT
RESPONSE TIME

FLAMESTAT DETECTS
FLAME AND OPERATES
FIRST CONTACT

IGNITION
PERIOD

IGNITOR OFF

IGNITION
OVER-RUN

PILOT OFF

Fig. 30 Hypothetical control for large burner with pilot

ignition period. The operator has to select a fuel feed rate which is greater than the heat demand rate in order to attain and maintain a relatively steady pressure or temperature. There is a tendency to overshoot the control limit, particularly with solid

118

fuel. For example when the control stops the fan (and fuel feed if an underfeed stoker), the fuel bed temperature drops slowly so that the pressure may rise above control setting before it starts to fall again. When the lower limit is reached a little time is required to regain fuelbed temperature before the fuelbed temperature rises sufficiently to stop the falling pressure.

This characteristic can be exaggerated if the boiler is oversized for the demand upon it and this often happens in heating installations during mild days in winter and at the start and end of the heating season. It is not unusual to see a burner on for five minutes and off for twenty minutes and occasionally off for as much as forty minutes. The boiler and associated flue and associated flue and chimney structure cool down and heat has to be provided to 'warm up' the system before effectively providing the required pressure or temperature conditions. This has an adverse effect on the overall thermal efficiency of the plant and is influenced by the ratio of time off to time on.

HIGH/LOW/OFF CONTROL

Some improvement on the foregoing characteristic swing and a narrowing of the cyclic band can be obtained by using a control with two switch positions. For example, as pressure or temperature rises, the first switch sends a signal to a device such as an electromagnet or a pulling motor which alters the fuel and air supplied by linkage to a low flow setting. On an underfeed stoker a small motor might be used to move the gear lever to a lower fuel feed position and the forced air damper to a partly closed position. With an oil or gas burner the fuel and air supplies are reduced together. When the set position is reached the system responds by shutting off fuel and air supplies. Thus the burner fluctuates between high and low with occasional 'off' periods during light load. The tendency to overshoot is reduced and fluctuations in temperature/pressure minimised. The tendency for smoke emission to occur due to the effect of cool surfaces chilling an oil flame is also minimised. The overall thermal efficiency is also improved by a reduction in the time-off/time-on ratio.

There is one advantage in an on/off system, since the fuel/air ratio is fixed and providing the burner, and its mechanical linkage between fuel and air remain without alteration, the system may be set for the highest practical CO_2 or lowest oxygen. In the case of a High/Low/Off and Modulating burner, the various burner, valve, air supply devices may not be capable of giving a constant CO_2 or O_2 content throughout the range of fuel inputs. Nevertheless these systems lead to higher overall thermal efficiency and steadier steam pressure and temperature.

MODULATING CONTROL

In order to maintain pressure or temperature within a minimum of fluctuation at the set point a proportioning band is introduced within which the control increases or decreases burning rate. With the simple On/Off or High/Low/Off reliance is upon the original manual setting of fuel/air ratio but with modulation fine adjustment of the ratio has to be achieved over the whole turn-down range by a profile cam. Further, in the larger boilers, particularly water-tube boilers, feedback systems are introduced so that draught conditions are progressively altered to suit load conditions. Some larger systems use rapid response oxygen analysers as trimming devices to alter the fuel/air ratio and of course the use of the so-called 'clean' fuels has made this system more effective since sample lines are less prone to blockage.

ANTICIPATORY/SEQUENTIAL CONTROL

By relating changes in steam flow to heat input a faster speed of response can be attained and yet more effective control achieved. A greater degree of flexibility is possible with multi-boiler installations where significant load changes occur throughout a given day.

EXAMPLE

Three boilers are available each at 11 363 kg/h steam output. The work load varies between 9090 kg/h and 31 850 kg/h. Clearly at 31 850 kg/h three boilers are required, whilst at 18 180 kg/h and 9090 kg/h. two boilers and one boiler, respectively, are required. By controlling from a master steam flow recorder, the system can be designed to use only the number of boilers related to the load to minimise heat and purging losses.

GENERAL

Water level controls of the on/off type can cause surging trouble when associated with automatic burners. If set to pump 10–25mm of water on demand, the weight of cool water may cause a relatively sudden pressure drop, accompanied by a rapid rise in fuel burning rate to re-attain the set pressure. The steam pressure suddenly rises when the water level is recovered and is often accompanied by a high firing rate. This is followed by a rapid change in firing rate to the low position and the cycle repeats itself. Where an on/off water level control is fitted it should be set to the minimum possible difference in water level. Ideally, fully modulating proportional type of control is recommended, where the feed control valve is progressively opened by changes in water level. Even with this system surging can occur if the control valve is incorrectly sized for the duty it has to perform.

Occasionally a steam flow meter will show a hunting characteristic which is sometimes erroneously blamed on changes in steam load and this may be due to the use of oversized pressure regulating or reducing sets. Similarly, violent periodic load changes could be due to a large steam user in the works being equipped with a simple On/Off temperature control.

INSTRUMENTS

The boiler house must be properly instrumented, not only to indicate the most suitable settings for the plant but to show the operator what happens as the result of his own actions.

DRAUGHT

A draught gauge, by indicating relatively slight degrees of suction or pressure, shows the effect of altering damper settings. The gauge also shows to what extent the suction in the flues and furnace is increased by widening the chimney damper opening. If the forced draught damper is opened up, furnace suction is reduced and a slight pressure may be created. The exact change is registered by the draught gauge, which also indicates if the flues or smoke tubes are blocked by deposits or dust.

The connection from the gauge to the flue should be robust, preferably of iron pipe, and the end should be at right angles to the direction of gas flow. A draught gauge should be clearly visible from the operating floor and the probe end cleared of any blockage as a routine duty.

CO_2 AND O_2

The air used in the combustion process should be completely under control, and a CO_2 (carbon dioxide) or O_2 (oxygen) indicator is necessary to determine that the correct amount of air is used. It also shows if unwanted air is being drawn through brickwork, holes in the fire, damper slots, inspection doors, or other places where it can detract from boiler efficiency.

Primary and secondary air needs to be adjusted to keep smoke emission within the requirements of the Clean Air Act while, at the same time, maintaining the highest percentage of CO_2 and the smallest percentage of O_2. There are various ways of measuring these gases in the flue gases. The chemical method involves the use of a portable apparatus, such as the Orsat shown in Fig. 31, which is easily used by an operator after instruction. This method is based on passing a sample of 100 volumes of flue gas through caustic soda or potash which absorbs the CO_2. The percentage of CO_2 in the flue gas is then indicated by the reduction in number of volumes from the original 100. The reduced volume is then treated with a solution of pyrogallol in caustic soda which absorbs the O_2. The second reduction in volume, but not the sum of both reductions, is the percentage of O_2. The remaining volume is then passed through a cuprous chloride solution to determine CO.

A simplified version of the Orsat can be used and Fig. 32 shows an example. This can be used for CO_2 or O_2 (not both) depending on the solution used.

Various electrical and electronic methods are now available for measuring CO_2 and O_2 and these are extremely simple to use. Some electronic instruments combine O_2 and temperature and do the stack-loss calculation automatically.

Other types of automatic instrument, suitable for mounting in the boiler house give a continuous indication of CO_2 on a dial or as a trace on a chart, and there is an increasing tendency to record O_2 content.

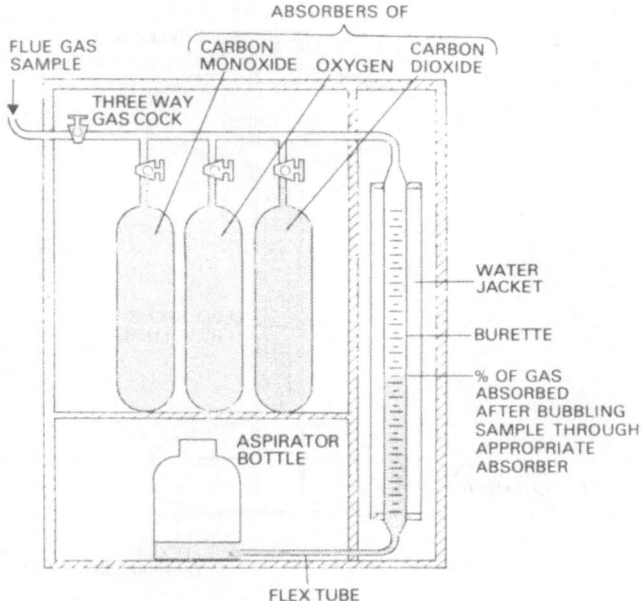

ABSORBERS OF

FLUE GAS
SAMPLE

CARBON
MONOXIDE   OXYGEN

CARBON
DIOXIDE

THREE WAY
GAS COCK

WATER
JACKET

BURETTE

% OF GAS
ABSORBED
AFTER BUBBLING
SAMPLE THROUGH
APPROPRIATE
ABSORBER

ASPIRATOR
BOTTLE

FLEX TUBE

Fig. 31 Orsat gas analyser

It is important to note that CO (carbon monoxide) in the waste gas, which in the case of coal or oil is usually accompanied by smoke, indicates a serious loss of efficiency.

All indicating and recording instruments need periodic checks against zero and a standard. Electrical equipment must be standardised on electrical potential. Obviously if a CO_2 meter is tested on 'air' the meter should register zero whilst in the case of an O_2 meter, the reading will be 21% and if the readings are any different there is clearly an error!

SMOKE DENSITY

A smoke density meter indicates the density of smoke in flues or chimney by measuring the degree to which light from a lamp passes through the flue gases to a detector. The lamp and cell housing require regular cleaning or a false impression of smoke

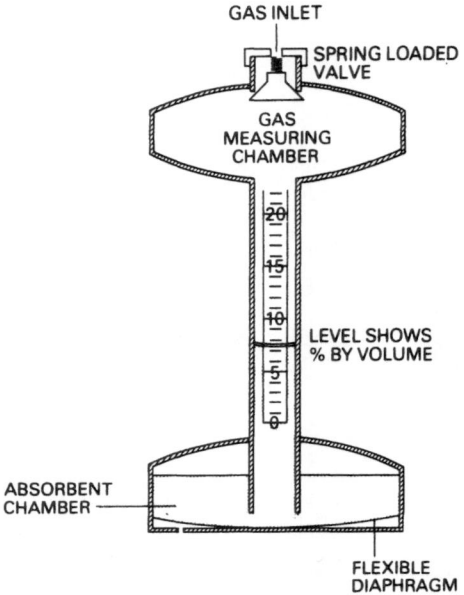

GAS INLET

SPRING LOADED
VALVE

GAS
MEASURING
CHAMBER

20

15

10

LEVEL SHOWS
% BY VOLUME

5

0

ABSORBENT
CHAMBER

FLEXIBLE
DIAPHRAGM

Fig. 32 Fyrite gas analyser

emission can be given. This meter is used to ensure that smoke is kept within the prescribed limits of the Clean Air Act.

TEMPERATURE

A temperature indicator or recorder shows the flue gas temperatures at various points in the boiler, flues and chimney, the combustion air temperature, the feedwater temperature into and out of the economiser and into the boiler, and the temperature of the steam leaving the superheater. This indicator is also needed to determine the oil temperature at the burner of an oil-fired boiler.

The CO_2 or O_2 reading and gas temperature at the boiler exit are used to discover how much heat is being wasted in the dry flue gases proceeding to the chimney (see Chapter 4).

If surfaces become fouled on the gas side or scaled on the water side, flue gas temperatures generally rise and water and steam temperatures fall. These temperatures should be logged by the operator so that an investigation can be made into significant changes, and the tube cleaning schedule properly established.

FUEL, WATER AND STEAM

Each boiler house should have some means of measuring fuel consumption, water evaporation, and the amount of steam produced. Coal consumption can be measured from recorded delivery or by calculating the volume of fuel in the bunker and using a carefully determined factor to obtain the weight. Some automatic proprietary devices work on the latter principle but need very careful and frequent calibration as well as recalibration if the fuel is changed.

Oil can be measured by 'dipping' a tank, provided care is taken to ensure the weight/volume factor is correct and adjusted to oil temperature. Oil meters are often installed and, if a strictly accurate test is being carried out, must be corrected for oil temperature.

The steam raised or water evaporated can be determined by a steam flow meter, which may also have a device for determining the total flow over a period. The meter must be correctly set for the pressure and condition of steam or the flow will need to be corrected by a factor. It must be 'blown down' periodically to ensure that the pressure tappings are all clear. A water meter for measuring boiler feed water can also be used to determine the amount of steam raised but allowance has to be made for blowdown or a change in the boiler water level. If strict accuracy is required, the water meter will also need correction for water temperature.

The efficiency of a boiler house depends on the amount of steam raised per unit weight of fuel. These figures should be logged, examined regularly and an investigation made if marked discrepancies are recorded.

ACCURACY

In all cases, the instrument maker's instruction book, which generally refers to a numbered instrument, should be kept carefully for reference. Generally, however, the maintaining of instruments in accurate working order is not the job of the boiler operator and he is strongly advised not to attempt work for which he has had neither the experience nor training. Doubtful equipment should be reported to the appropriate authority so that an instrument mechanic can carry out repairs or make the necessary adjustments. Simple routine maintenance is generally all that is required.

Chapter 13

SAFETY

STATUTORY SAFETY EQUIPMENT FOR BOILER PLANT

On matters of Safety the operator must remember that he must follow the safety rules laid down by his employer. Boiler plants vary enormously in design and installation and it is not possible to lay down a rigid set of rules covering every type. The following is a brief resumé of the essential requirements only.

A steam boiler is capable of causing destruction and even death if it is not correctly maintained and operated.

The Factories Act 1961, with subsequent revisions, and now incorporated in the Health and Safety at Work Act, lays down specific rules to be observed by the owner of the plant, and the boiler operator. Many operators attending NIFES courses have never seen appropriate sections of the Act and the readers of this book are urged to obtain a copy of the Act relating to boilers, in particular Technical Data Note 25 *Safe operation of automatically controlled steam and hot water boilers* (H.M. Factory Inspectorate) published by HMSO, and a copy of the *Requirements for Automatically Controlled Steam Boilers* issued by the Associated Offices Technical Committee, Longridge House, Manchester M60 4DT.

The following are extracts from Regulations but it is **urged yet again** that the appropriate sections of the Act be read in detail.

Every steam boiler whether separate or one of a range shall have attached to it:

    (a) A suitable safety valve separate from any stop valve, which shall be adjusted to prevent the boiler being worked at a pressure greater than the maximum permissible working pressure and shall be fixed directly to or as close as practicable to the boiler.

NOTE: A lever valve shall not be deemed a suitable safety valve unless the weight is secured on the lever in the correct position.

(b) A suitable stop valve connecting the boiler to the steam pipe. Where two or more boilers feed a common steam pipe serving a process requiring a continuous supply of steam, a device to enable the stop valve to be isolated for examination purposes must be provided.

(c) A correct steam pressure gauge connected to the steam space and easily visible by the boiler attendant, which shall indicate the pressure of steam in the boiler in the recognised unit and have marked on it in a distinctive colour, the maximum permissible working pressure.

NOTE: The maximum permissible pressure is set by the Insurance Company and must not be revised without reference to the Insurance Company.

Most shell-type boilers have the pressure gauge visible from the firing floor but with larger water-tube boilers, the gauge reading may have to be relayed from the steam space at the top of the boiler to a suitable position on the firing floor and allowance made for the static head of water in the connecting pipe.

(d) At least one water gauge of transparent material or other type approved by the Chief Inspector to show the water level in the boiler together with, if the gauge is of the glass tubular type and the working pressure of the boiler normally exceeds 2.75 bar, an efficient guard provided so as not to obstruct the reading on the gauge.

(e) Where the plant is of two or more boilers, a plate bearing a distinctive number which shall be easily visible.

(f) A means for attaching a test pressure gauge.

(g) Unless externally fired, a boiler must be provided with a fusible plug, or an efficient low water alarm device.

NOTE: A fusible plug shown in Fig. 33 was fitted to the older boilers and in sight of the maximum heat source, usually an incandescent coal bed. With the advent of oil and gas firing method and especially on 'Packaged' boilers, the fusible plug is no longer incorporated and reliance is placed entirely on automatic low water cut out.

BOILER SAFETY

It is the operator's duty to ensure that the boiler is operated efficiently and safely. He must see that the water in the boiler is kept at the correct level and that the boiler does not become overheated.

Frequent observations of the gauge glass is one of the most important functions of the boiler attendant's duties. It is the means by which failure of automatic control systems can be detected and therefore the attendant must ensure that the gauge glasses indicate correctly by adherence to gauge glass drill, which will be described later.

The fusible plug mentioned previously prevents the risk of an explosion by the melting of a central bore in the plug when it is uncovered through the water level falling. Steam will escape through the hole in the plug and give timely warning of low water as well as quenching the fire. The surface of the plug on the furnace side should be kept free from deposits. (See Fig. 33.) Every steam boiler and all its fittings and attachments shall be properly maintained.

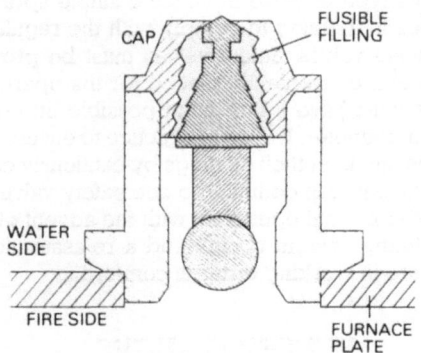

Fig. 33 Fusible plug

No new steam boiler shall be taken into use unless there has been obtained from the manufacturer of the boiler, or from a boiler inspecting company or association, a certificate specifying its maximum permissible working pressure and stating the nature of the test to which the boiler and fittings have been submitted, and the certificate is kept available for inspection and the boiler is so marked so as to enable it to be identified as the boiler to which the certificate relates.

The maximum time allowed by the Act, between each official inspection and report is normally fourteen months.

NOTES ON SAFETY VALVES ARISING FROM INSURANCE CO. REQUIREMENTS FOR STEAM BOILERS

Detailed regulations cover the size of valves in relation to boiler capacity and a detailed description is beyond the scope of this handbook.

The main requirements are that every boiler should have two safety valves each alone capable of dealing with the maximum rate of steam production. Valves must not be less than two inches in diameter and must be mounted directly on the shell or steam drum.

The locomotive type of valve in which a single spring controls a pair of valves is deemed to comply with the regulations concerning duplicate valves. Safety valves must be provided with hand easing gear, conveniently placed for the operator. Spring loaded valves must have a maximum possible lift equal to one quarter of their diameter. It is good practice to ensure that valves do not become stuck on their seatings by cautiously easing once per week. Whilst it was common to see safety valves lifting or feathering under normal operation, with the advent of automatic control such lifting seldom occurs and a re-assurance that the safety valves are in working order is comforting.

WATER GAUGES

It is desirable for two completely independent water gauges to be provided on the front of the shell or drum. The practice which is often adopted of having two gauges on a single mounting (to reduce the number of holes cut in the boiler shell) is not ideal as both gauges can be put out of action or rendered unreliable by a blockage in the common connection.

In fact it is not unusual to see a pressure gauge fitted to the same mounting and there are instances where tappings are taken from the same single mounting to audible alarm systems. Gauge glasses should be protected by a frame with sides and front made of heavy plate glass. A white backing plate with diagonal black stripes is useful, as the refraction caused by the water in the glass causes the black stripes to distort visibly and the water level can be readily seen.

Before removing guards for daily cleaning, the water and steam cocks should be closed and the drain cock opened.

Gauge glasses can fail very rapidly by a thinning of the glass due to a high alkalinity in the boiler water. Gauge glasses and guards should always be kept clean, and be tested by blowing down every shift where continuous working is carried out, and at least once per day, where boilers are steamed for less than twenty four hours per day.

GAUGE GLASS DRILL

The gauges should be blown down every shift where continuous working is carried out, and at least once per day, where boilers are steamed for less than twenty four hours per day.

An accepted procedure for blowing down gauge glasses is as follows:
    (1) Close water cock.
    (2) Open drain cock (note that steam escapes freely).
    (3) Close drain cock.
    (4) Close steam cock.
    (5) Open water cock.
    (6) Open drain cock (note that water escapes freely).
    (7) Close drain cock.
    (8) Open steam cock.
    (9) Open and then close drain cock for final blow-through.

The water level should be quickly restored in the gauge glass, the water that first appears is generally representative of the boiler water and if discoloured the reason should be ascertained. Normally there should be no discoloration except in cases where internal treatment of the boiler water is used.

Drains from gauge glasses should be safely led to an open

tundish or drain, where the drained water can be seen to be flowing freely. When blowing down gauges caution must be observed to prevent scalding by water splashing from the tundish, and a suitable guard may be necessary to prevent this.

If the water level in the gauge glass fluctuates violently, priming is probably the reason. A definition of priming has already been given in the section on water treatment.

AUTOMATICALLY CONTROLLED STEAM AND HOT WATER BOILERS

It is worth quoting from a report published by one of the Engineering Insurance Groups.

'Attendants of steam boilers are in fact custodians of potential bombs, but so many have become so used to the apparent docility of their charges that nonchalance has taken the place of a one time healthy respect.'

The same report analyses the various reasons for accidents but the report positively stressed that none of the accidents investigated should have happened if proper attention had been paid to the water gauges and water level controls.

At one time automatic equipment was installed to control combustion and feedwater in relation to changes in steam demand, but continuous attendance was maintained by an operator who had the ultimate responsibility of shutting down a boiler which he considered to be unsafe.

Modern trends however have been to dispense with regular attendance, and control equipment has been designed with this in mind. Even so, an increase in accidents has been reported as being due to the non-functioning of water controls and to lack of knowledge or indifference to carrying out tests and maintenance. Even the casual boiler attendant has considerable responsibility in testing and checking controls and for emergency shut down procedures. The requirements for automatically controlled steam and hot water boilers can be read in publication Technical Data Note 25 (H.M. Facotry Inspectorate), and the publication states, in relation to feed water supply, that for boilers not continuously attended the automatic level controls must positively control boiler feedpumps or regulate water supply to maintain the water level within certain predetermined limits.

When the water level falls to a predetermined level, an audible alarm should be simultaneously operated, the burner will cut out; however, restoration of the water level to normal working level may allow an automatic burner to be restarted at this stage.

When the water level falls to a predetermined level below the first predetermined level, deemed extra low water, an independent overriding control must cut off oil or gas supply, or air supply in the case of a solid fuel fired boiler. An audible alarm must sound, and the control must lock out, and be of the type that requires manual resetting before the boiler can be brought back into operation.

All electrical equipment for water level and firing controls should be so designed that faults in the circuit will automatically shut off fuel and air supplies to the boiler.

BLOWING DOWN OF WATER LEVEL CONTROLS

Steam and water legs and control chambers should be blown down at least once every eight hours of normal steaming to ensure that they are clear and that fuel or air cut-outs are working properly.

Isolation of the control chamber caused by boiler attendants closing, and leaving closed, either steam or water isolating valves or both, after closing the drain valve has been the cause of many boiler explosions, resulting from overheating caused by low water.

To prevent isolation of control chambers, sequencing blowdown valves have been developed and these are now invariably fitted to modern package boilers.

The sequencing blowdown valve is arranged with a series of ports, so that the water leg is not isolated without the drain being open. The float will, therefore, be at the bottom of the chamber so automatically cutting off the fuel supply to the burner, and making it impossible to relight under these conditions. The arrangement of the valve ports enables the steam and water connections to be blown through separately. The valve is also arranged with a ratchet so that once the operation of blowing down has been commenced the valve cannot be closed until the full sequence has been completed. A further advantage of this valve is that the possibility of damage to the float resulting from water violently entering the float chamber is averted.

In some cases boilers cannot be shut down to enable maintenance to be carried out on control chambers, and in these cases it is necessary to fit steam isolating valves. When this is so, the valves should be locked (under normal operation) in the open position, and the keys kept by a responsible person. A duplicate key should, however, be available in a glass fronted case for emergency use.

OTHER SAFETY ASPECTS

BOILER BLOWDOWN

The instructions issued by the boiler manufacturer frequently include blowdown procedure and these should be studied carefully as blowdown can sometimes seriously interfere with water circulation over the heating surfaces and cause damage. A routine blowdown procedure should be established and maintained. When using the main blow-down cock the operator must not, under any circumstances, leave the valve unattended until the blow-down procedure has been completed. If more than one boiler is connected to a common blow-down system, even if the boilers are not in the same boiler house, the law requires that only one blowdown key for that system shall be available in the works.

This applies even if the boilers are of different sizes and have separate blowdown connections into the system. In the latter case, the valves have to be standardised or if there are two boilers with different sizes of key, it is permissible to weld the keys end to end and make one tool of them. If the valve is not fully shut, the blowdown key cannot be removed from it but the key should not, in any case, be left in a closed valve.

Where one or more boilers on a steam range are out of commission and open for cleaning, the blowdown must be blanked off to avoid blowdown from another boiler passing back through a leaking valve. Care must always be taken to ensure that an open boiler is safe before another boiler on the steam range is blown down.

PRECAUTIONS REGARDING WORK

PREPARATION FOR INSPECTION AND MAINTENANCE
After isolating the boiler (steam, water and fuel connections) it should be allowed to cool slowly.

Rapid cooling hardens scale in the boiler and may cause joints to open. When zero pressure shows on the gauge, test or air cocks must be opened and safety valves lifted where possible, and the water may be run out. It is important that any manhole or access cover should not be removed or opened, whilst there is the slightest possibility of any pressure or vacuum condition existing in the boiler. Reliance should not be placed entirely upon the pressure gauge and sufficient margin of time must be allowed after the pressure has fallen to zero and after test or air release cocks are opened. Before entering a boiler a check must be made to verify that all steam, water and blowdown valves are tightly closed. Where the boiler is connected to a range, both boiler crown valve and range isolating valves and feed check valve should be locked with padlock and chain. Alternatively, connections must be blanked or spaded off. Many serious accidents have occurred owing to steam or water being admitted to boilers in which men have been working. The Regulations state that where a boiler is one of a range of two or more, no persons shall enter or be in any steam boiler unless: All inlets through which steam or hot water might otherwise enter the boiler from any part of the range are disconnected from that part; or, all valves or taps controlling the entry of steam or hotwater are closed and securely locked.

Notices should be placed at clearly visible locations to warn that men are working inside the boiler and a responsible person must be aware that men are inside the boiler.

GENERAL SAFETY POINTS

Cases have been known, where a blowdown valve has choked up with scale, for the tail pipe to be removed and attempts made to free the blockage by rodding whilst under pressure. Needless to say, fatalities have occurred and under no circumstances must this be attempted.

Care should be taken when breaking joints on steam mains in case isolating valves are not tight.

Catwalks should be uncluttered; tools, bits and pieces of removed material can fall on people below. Such debris on the catwalk can cause people to trip resulting in broken limbs and sometimes death. Safety helmets should be worn.

Spilt oil should not accumulate since it can cause serious accident by persons slipping and of course presents a potential fire hazard.

Make sure that all tools are removed from inside a boiler before boxing up.

Many boiler houses are inadequately lit, if this is the case ask for more lighting and make sure that the fittings are subsequently kept in a clean condition.

An attendant should always use a blue or green viewing glass screen or goggles for eye protection when looking at the flame on fuel bed through the inspection port. If the boiler house is very noisy and nothing can be done to abate the nuisance at source, ear muffles can be worn.

Adequate ventilation is necessary to prevent asphyxiation within the boiler house; do not seal up established openings and do not aggravate the problem by allowing the flue gas passages to foul up so that fumes enter the boiler house. Very many deaths have occurred in this way, particularly in boiler houses below ground level that are looked after by caretakers having other duties to perform.

Enter flues with caution and notify a responsible person that work inside the flues is proceeding. Put up a notice that work inside is in progress. Wear breathing apparatus unless the area has been declared safe for working. Request that the flues be examined and declared devoid of personnel before closing up.

OIL STORAGE

Do not enter oil storage tanks without first seeking Authority for clearance nor before ascertaining the depth of sludge in the tank; or that the tank is free from toxic gases and explosive vapours. Anyone inside a storage tank should be in full view of someone outside and a workman who develops a headache or finds his vision affected should warn his companions and get into the fresh air as quickly as possible. The matter should be reported immediately to a responsible person.

Obey No Smoking Regulations in the vicinity of oil tanks. Naked flames should not be used. Bulk oil is usually stored in tanks surrounded by a bund wall to prevent oil from a burst or leaking tank from spreading and so becoming a fire risk. The bund volume must be adequate to contain the whole contents of the tank and should always be kept clear of rubbish. The valve on the outlet of the tank is often kept open by a fusible link, which melts when it becomes hot and closes the oil valve. A fire extinguisher or sand should be used to tackle an oil fire and some installations are fitted with automatic fire extinguishers. In no circumstances should water be used.

BURNERS AND COMBUSTION

The operator must verify that no person is in the flue system before lighting up. He should ensure that the boiler outlet damper is open and also any other damper which, if shut, would impede the flow of gases to atmosphere. If a fully automatic boiler is used it should incorporate a lock out in the system which will not permit flame establishment if the damper is shut. In no way should this lock out be circumvented.

In a non-automatic boiler it should be arranged that the flue damper cannot be fully shut. Lock out on failure of either forced draught or induced draught fans is required on fully automatic boilers.

For boilers not continuously supervised automatic firing controls must be arranged so that manual re-setting is required on lock out due to flame or pilot flame failure on oil and gas boilers. A lock out having manual re-setting is required if the fuel does not ignite within a pre-determined time.

The flame detector device should be checked regularly and properly maintained.

Lock out should occur with manual re-setting when the excess pressure- or temperaturestat comes into operation.

The cause of the failure to cut out on the normal operating stats should be ascertained.

EXPLOSION IN THE FLUES

If raw or incompletely burned gases or fuel are mixed with air and heated, a blowback of flames from the front of the boiler, or even an explosion can result. The same type of danger is possible when a boiler fired with solid fuel is left banked. Similarly, if the flame goes out when using oil, gas or PF explosive air/fuel mixtures can accumulate. The remedy has already been described in the previous paragraphs, but the action **purge** is re-emphasised.

ECONOMISER SAFETY

If the water temperature becomes too high or the water supply fails, the economiser must receive immediate attention. The boiler itself will be safe until the water falls to the alarm level, so that there will be sufficient time to by-pass the economiser, draw-off the fires, or if manually operated, turn off the burners, before this happens. Economiser explosions can be very dangerous.

USE OF GAS

When a gaseous fuel is used, under no circumstances must naked flame be used for checking gas leaks. Any suspicion of gas leakage should be reported to a responsible person so that 'authorised' inspection can be carried out. Special equipment including gas detectors and, in confined areas, respirators is needed. If, however, an operator experiences headache, sickness or giddiness, he should warn anyone in the vicinity and get into fresh air at once.The matter must be reported to a responsible person.

It is very important to remember that LPG, unlike most other gaseous fuels, is heavier than air and tends to accumulate in low or submerged places, e.g. below ground level and in basement boiler houses.

COAL STORAGE

A coal stack can heat up and catch fire by spontaneous combustion unless it is sufficiently ventilated to allow air to carry away the hot gases or is sealed with fine coal to exclude the air. Coal must not be stacked close to the boiler flues or built higher than 2m. Individual stores of more than 200 tonne are inadvisable.

If a stack ignites, it must be broken into to find the seat of the fire and the hot coal quickly moved away from the rest and scattered on the ground. The stack should not be re-built until all the hot coal has been removed.

COAL AND THE BOILER HOUSE

It is difficult to avoid fine dust accumulations, especially at high level boiler and boiler house support members. If a fire occurs at these points it should be smothered and as far as possible not disturbed.

Explosions can happen if the dust is blown forward by a high pressure water hose and fatalities have occurred.

CONCLUSION

It is reiterated that all the relevant safety regulations should be studied by boiler attendants already in post and those who intend taking up duties as such.

Safety is now very well covered by the City and Guilds of London Institution in their Boiler Operators Certificate course and boiler attendants will derive great benefit from this course and incidentally derive much greater job satisfaction.

APATHY is the real **ENEMY**. Education and enlightenment is the surest route to safety.

Chapter 14

THE CLEAN AIR ACT

The first Act of Parliament to deal with air pollution was passed in 1956. Some of the sections were concerned with smoke from domestic chimneys and others ensured that new furnaces and boiler plant are properly planned. The Act made it an offence to allow smoke or grit beyond a specified amount to be discharged from any industrial chimneys and the boiler operator must therefore know how to avoid infringing the law. The 1956 Act has been reviewed and a second Clean Air Act was passed in 1968.

SMOKE

The Act distinguishes between what is called dark smoke and black smoke, the colour being determined by comparison with a Ringelmann Chart, a miniature of which is shown in Fig 34. If this chart is propped up across the room the white squares fade out and the card looks grey or black according to the thickness of the lines. The standard chart is of course much bigger and is viewed at a distance of 15.2m.

The Act stated that dark smoke is as dark as Chart 2 and black smoke as Chart 4. It is an offence to emit dark smoke from any chimney in normal operation but allowances are made for lighting up, soot blowing, and flue cleaning. With a single boiler, dark smoke is allowed for these purposes for up to 10 minutes in a period of 8 hours and can be split into five periods of 2 minutes or four periods of 2½ minutes, or anything which adds up to 10, provided that no single period is longer than 4 minutes.

When a soot blower is used, the limit of 4 minutes for any single period of smoke is relaxed and the total allowance is increased to 14 minutes in 8 hours.

Number of boilers	ALLOWANCE	
	Without Soot Blowing min.	With Soot Blowing min.
1	10	14
2	18	25
3	24	34
4 or more	29	41

If more than one boiler feeds into the same stack, the allowance in 8 hours is slightly more.

Black smoke, Shade 4, is treated far more seriously because it indicates extremely bad combustion conditions. The limit is 2 minutes in any 30 minutes which, in effect, means that no black smoke can be discharged under normal working conditions. An exception is made for a fire being lit with coal but, even then, the smoke must be reduced as quickly as possible and there is no separate allowance over and above the periods permitted for dark smoke.

SMOKE CONTROL AREAS

In the smoke control areas now established in the large cities discharge of smoke is prohibited and no allowances are made.

Smoke can be avoided by using smokeless fuels, such as coke or anthracite, and a list of fuels authorised for smokeless zones, which includes low temperature coke, and low volatile steam coals, has been issued. Exemption may also be given for using other solid fuels, particularly in mechanical stokers; liquid fuels can be used provided smoke emission is kept to a minimum.

GRIT EMISSION AND DISPERSION OF COMBUSTION PRODUCTS

Grit is defined as being solids over 0.075mm or 75 microns whilst dust is considered to be less than 75 microns. The 1968 Clean Air Act states:

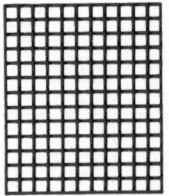

SHADE 1
DENSITY 20%

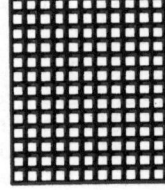

SHADE 2
DENSITY 40%

SHADE 3
DENSITY 60%

SHADE 4
DENSITY 80%

Fig. 34 Ringelmann chart

142

(i) That plant burning more than 45.45 kg solid fuel per hour must have gas cleaning equipment and that prior approval of the plant must be given by the Local Authority.

(ii) The chimney must also be approved by the Local Authority. The Local Authority assesses chimney height according to the Memorandum of Chimney Heights, published by the Stationary Office.

(iii) The legal limits for grit and dust are contained in SI 1971 No. 162 obtainable from HMSO.
The 1971 regulations applied in the November of that year for new plant and apply from 1st January 1978 for existing plant. Regulations on the measurement of grit and dust are contained in SI 1971 No. 161.

CHIMNEY HEIGHT

The chimney height is related to the weight of fuel burned and its sulphur content. Recommendations are also made regarding the efflux velocity of the flue gases which for large chimneys may be as high as 15.24 m/s at full load for boilers rated at 204 545 kg/h of steam and not less than 7.62 m/s for boilers rated up to 13 636 kg/h of steam. The Memorandum relates the height to the type of surrounding district ranging from an underdeveloped area having no background pollution to a large city with severe background pollution. Five areas of differing characters are given in the Memorandum.

It is not too difficult to apply design parameters to new plant but existing plant could present a problem when it comes to securing effective dispersal of sulphur dioxide and fine dust into the atmosphere, such as not to make the concentration or fall out at ground level offensive to the populace. The maximum ground level concentration of gases e.g. sulphur dioxide, resulting from the burning of sulphur in the fuel and the maximum deposition of the finer dust is usually proportional to the height at which the nuisance starts to disperse.

This height is composed of the actual height of the chimney and the height of the plume issuing from the chimney. The latter height is governed by the buoyancy of the plume and its velocity. Therefore, the mass of the gas and its temperature are very important factors in securing effective dispersal. The requirement of stated efflux velocities in the Memorandum can now be seen in the correct perspective.

These factors are outside the control of the boiler attendant in as much as they are design parameters, but there are steps he can take personally or refer his observations to the Management to secure more effective dispersal.

(i) Do the flues and chimney need insulation to maintain as high a flue gas temperature as possible, consistent with the load and thermal efficiency of the boiler.

(ii) He should report any sections of insulation not replaced after repairs have been carried out.

(iii) He should look for any point where cold air can leak into the flue system thereby reducing the temperature of the flue gases. Inspection covers, overlapping but unsealed casing points, idle boilers are common sources of cold air infiltration.

(iv) Boilers operating at light load can result in a nuisance near to the source of the plant, since the efflux velocity and the plume rise are reduced in relation to the very much smaller weight of fuel consumed. This condition is often prevalent in those establishments having a high winter space heating load and a low process load which prevails all the year round. The operator should study his plant operating conditions so that he can assess, for example, whether he needs to use two boilers on light load or one boiler at a load near to its design capacity. With two lightly loaded boilers, burners could have long off periods, when the whole flue and chimney systems are coiling down. The resultant flue gases would have a much lower mean temperature than that of a boiler operating at high load factor. Investigations by NIFES Engineers have shown that many cases of smut emission are during the summer period rather than during the winter.

ACID SMUT FORMATION

Acid problems are not entirely associated with burning oil, they can also occur when burning highly volatile high sulphur coals.

The solid particles emitted from oil burners consist mainly of fine carbon which may contain fuel ash and sometimes a small proportion of bitumen-like material. The name 'Cenosphere' is given to these particles, which are roughly spherical in shape and hollow, hence the term. Their size may vary from a few microns up to about 200 microns outside diameter and depends mainly upon the quality of atomisation. Therefore, it can be seen that the effective way of reducing solids with oil burning equipment is one of securing good atomisation with the existing plant, by ensuring that operating technique is satisfactory, e.g. minimum excess air, avoidance of flame impingement and above all, the care and maintenance of the burner. Having satisfied oneself that operating conditions are satisfactory but that a nuisance still prevails, then compliance with solids emission standards is a matter of replacing the burner system rather than the provision of arresting equipment.

With good atomisation and combustion, few cenospheres will exceed say 30 microns. Sometimes individual particles may bind together or agglomerate to form loose clusters in suspension in the gas stream and around which sulphuric acid (SO_3) can 'nucleate'. We now have the condition known as smutting. Smuts are usually the result of poor combustion which produces a high concentration of total stack solids in combination with SO_3. Similarly agglomerates may be produced when there is flame impingement onto relatively cold surfaces where sulphuric acid may form. Occasionally, the fine particles in a flue stream will adhere to a relatively cool surface and flake away under changes of draught/temperature and/or rate of gas flow. One can seldom see the smuts discharge from the chimney and this is because of their relatively small number and large size they do not obstruct the passage of light, unlike smoke which is composed of a great number of very minute particles which absorb light and make the plume opaque or dark. Smuts, on falling, generally flatten under impact and produce the characteristic black greasy stain which is acidic and can cause damage.

It is a mistake to think that because a relatively high flue gas temperature prevails, that a flue or stack need not be insulated.

Tests were due to be carried out on a boiler and the flue gas temperature was known to be about 301°C apparently well above the acid dewpoint. It was requested that gas sample points be provided and whilst showing the desired location of such points with a pencil, the latter penetrated the casing, which was found to be badly corroded. The temperature in close proximity to the metal surface was only 65°C. An extreme case perhaps, but it re-emphasises the need to consider the insulation of flues.

The factors which need to be considered when minimising the extent of smut emission are the same as those which are required to give the maximum buoyancy effect to flue gases to give the most effective dispersal of SO_2 and fine dust.

It is possible of course, even with a well run boiler plant and a well insulated flue and stack system that load is the main factor influencing both smut formation and subsequent dispersal. If the problem still persists after having exhausted the possibility of reducing the number of boilers on line then fuel oil additives should be considered, especially those compounded to fix the sulphur oxides, especially the trioxide. The additive may need to be applied with the liquid fuel and injected as liquid or in powder form.

'Chinese Hats' a shallow cone held by spacers a small distance above the top of the chimney should be avoided, since they impede the ejection of flue gases and stack solids into the atmosphere and encourage the downwash of the plume near to the plant. Because of the cooling effect they also corrode quickly.

If it is the intention to minimise the effect of rain entering the stack during the idle periods then a sloping plate can be inserted in the chimney base and having a drain pipe which allows any water to drain away. An empirical measure of solids content in the flue gas from an oil fired boiler can be assessed from the determination of the Bacharach Number. A known volume of flue gases is drawn through a filter disc and varying shades of stain are produced dependent upon the extent of the concentration. The shaded areas are compared with a master scale numbered 1 to 9.

GRIT ARRESTOR

In this context, the combustion of solid fuel is being considered. During combustion ash is disposed of in several ways.

(i) The large agglomerates range from fused lumps to the relatively small particles falling through the bars into the ashpit or carried forward into the ash pit by some form of moving grate.

(ii) Deposit on internal surfaces, and

(iii) small particles entrained in the flue gas stream, some to be rejected into the atmosphere.

The solid particles will contain some unburnt carbon and the extent of the emission will be influenced by the rank of coal used, its ash content and the size range of the coal in use. A coal with a high proportion of fines is likely to lead to a high emission rate compared with one having a lesser proportion. Some coals, notably semi-anthracites, decrepitate under the action of heat and eject solid particles into the gas stream.

Operational factors influence the extent of emission. A high rate of burning will demand a higher air flow rate and velocity through the bars or links. This velocity may be sufficient to lift off the fine particles from the bed where they are entrained in the flue gas stream. Some of the solids deposit on the boiler surfaces to become an operating nuisance in their own right, since they impede flue gas flow and reduce the heat transfer rate.

A proportion will deposit in the flues especially at points of directional change or where a fall in flue gas velocity is likely. This is the principle of settling chambers and as a form of arresting plant may be sufficient in some cases to reduce the total emission to an acceptable level.

The factors governing subsequent dispersal and ground level concentration have already been discussed.

It is imperative for an operator to know his coal and its behaviour on the plant in relation to the load pattern. For example, if a coal is too dry the simple expedient of wetting or steaming in the hoppers or shutes could reduce the extent of emissions. Segregation of coal in hoppers which causes uneven distribution across the grates is a factor which can considerably increase emission.

There is a greater danger of grit emission with some types of mechanical stoker than others.

Generally, however, the important factor is the rate at which flue gases pass through the boiler plant. Reducing excess air and draught to a minimum reduces the velocity of gases and they

147

have less carrying power for grit, and the nuisance is minimised, or the work to be done by a collecting device reduced.

If the fuel bed is disturbed too much, fine coal and ash are lifted into the gas stream and cause grit emission. If the flues are dirty, dust deposits tend to get picked up, particularly when the draught has to be increased to meet a higher load. This compares with the breaking off of soot flakes in oil fired boilers.

In addition to settling or expansion chambers which may or may not be provided with a water curtain, other methods of separating dust from flue gases are:

(a) Cyclone or centrifugal collectors.
(b) Electrostatic precipitators.
(c) Bag Filters.

and sometimes combinations of these systems.

Simple types of grit arrestor are shown in Figs 35 and 36. The flue gases entering the arrestor are deflected by metal cones and expanded into a larger chamber which slows them down. The grit particles then fall out, strike the sloping plate, and slide down into the collector. Clean gas leaves by the central exit.

A general scheme of an electrostatic precipitator is shown in Fig. 37.

The solid particles are given a negative electric charge as the flue gases pass through earthed tubes or plates. The charged particles are attracted to the positively earthed collecting electrodes and when these are rapped, result in the detachment of the particles and deposition in the collection hoppers. It must be borne in mind that fly ash, particularly that from a pulverised fuel fired plant, is extremely abrasive and considerable erosion of hoppers and discharge valves is possible. Often, surfaces are fitted with wearing liners and it is possible for rotary valves to be subjected to considerable erosion effects.

All arrestors need valved grit and dust outlets to ensure that cold air is not sucked into the system since this could cause the flue and gas temperature to fall below acid dewpoint creating corrosive conditions, and causing the flue dust to 'cement'.

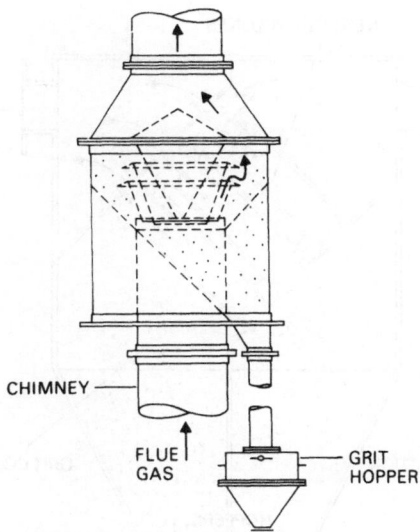

Fig. 35 Simple grit arrestor

EFFICIENCY OF GRIT ARRESTORS

The collecting efficiency depends upon its design, the grit and dust content of the flue gases, the particle size range of the solids, the mass flow of gases through the system.

SIMPLE INERTIAL TYPE
Collecting efficiency 70% for particles 70 to 80 micron size, but lower as particle size falls below 70 microns.

MULTI-CYCLONE UNIT
Collecting efficiency up to 90% for particles as small as 5 microns.

ELECTROSTATIC PRECIPITATORS
Typical units may have a collecting efficiency of 98 to 99% over all particle sizes and can eliminate dust and smoke particles down to 1 micron or less.

FLUIDISED BED
The future might see the boiler attendant looking after fluidised bed fired units and already it is claimed that such furnaces can

149

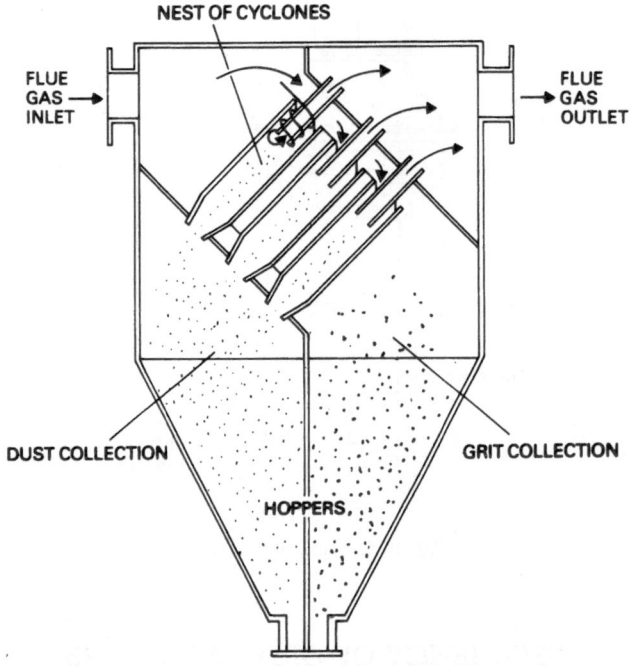

FLUE GAS INLET →

NEST OF CYCLONES

FLUE GAS OUTLET →

DUST COLLECTION

GRIT COLLECTION

HOPPERS.

Fig. 36 Multicell grit arrestor

burn a wider range of fuels with significant reductions in sulphur dioxide and nitrous oxide emissions.

Significant reductions in solid emissions are also claimed and one company claims that their fluidised bed fired boiler has not had to have a major cleaning in two years.

Clearly a reduction in solids emission might mean less elaborate auxiliary equipment in the shape of grit arresting equipment and a lowering of the overall capital cost associated with a coal fired plant.

A more detailed account of the application of fluid-bed combustion must be left to a future edition when the technique has been thoroughly established.

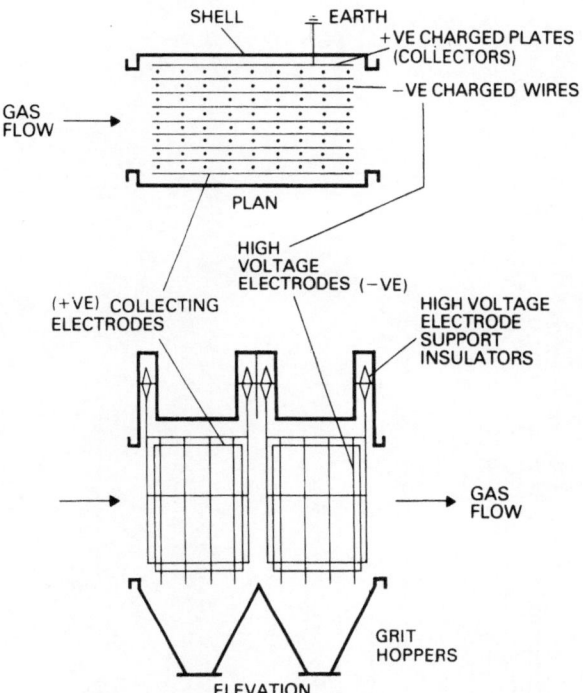

SHELL ___ EARTH
+ VE CHARGED PLATES
(COLLECTORS)
− VE CHARGED WIRES

GAS FLOW →

PLAN

(+VE) COLLECTING ELECTRODES

HIGH VOLTAGE ELECTRODES (−VE)

HIGH VOLTAGE ELECTRODE SUPPORT INSULATORS

→ GAS FLOW

GRIT HOPPERS

ELEVATION

Fig. 37 General scheme of electrostatic precipitator